SOLAR ARCHITECTURE

Proceedings of the Aspen Energy Forum 1977
Aspen, Colorado

Editors: **Gregory E. Franta, AIA**
Kenneth R. Olson

ROARING FORK RESOURCE CENTER

Production Coordinator: Heidi Hoffmann
Graphics: T. Michael Manchester
Proceedings Document Preparation: Central Transfer Company
Cover Design: David Brownell

ANN ARBOR SCIENCE
PUBLISHERS INC
P.O. BOX 1425 • ANN ARBOR, MICH. 48106

Third Printing, 1979
Second Printing, 1978

Copyright © 1978 by Ann Arbor Science Publishers, Inc.
P.O. Box 1425, Ann Arbor, Michigan 48106

Library of Congress Catalog Card Number 77-91356
ISBN 0-250-40233-5

Manufactured in the United States of America
All Rights Reserved

PREFACE

The Aspen Energy Forum 1977 was the fourth annual energy forum sponsored and coordinated by the Roaring Fork Resource Center of Aspen, Colorado. The theme for this forum was "Solar Architecture," emphasizing passive solar design. The forum was held on May 27, 28 and 29, 1977 at the Aspen Institute for Humanistic Studies, Aspen, Colorado. The facilities for the forum were filled to capacity with over 500 participants for the three day event. Approximately 37% of the participants were local residents in the Roaring Fork Valley, 34% were from Colorado (non-local) and 29% were from outside the State of Colorado.

Thirty-five speakers from throughout the country presented the objectives of the Aspen Energy Forum 1977, which were:

1. To enhance awareness of the magnitude of the global energy and resource enigma to the design profession and the general public.
2. To provide an introduction, design tools and an overall comprehension of the architectural technologies of passive solar systems.
3. To provide fundamentals in building design in relation to climate, energy and resource conservation and alternate energy sources.
4. To stimulate awareness of the benefits of food and heat producing solar greenhouses.
5. To illustrate applications of energy and environmental education through case studies.

The Proceedings of the Aspen Energy Forum 1977 contained in this document include papers and transcriptions of the presentations made at the Forum. The content of the proceedings are opinions of the presenters and do not necessarily reflect the views of the Roaring Fork Resource Center. However, the Roaring Fork Resource Center has compiled, edited and prepared these documents, and correspondence regarding the content may be addressed to:

> Roaring Fork Resource Center
> P.O. Box 9950
> Aspen, Colorado 81611   303/925-8885

The Roaring Fork Resource Center is a non-profit educational organization dedicated to the preservation of natural resources through the use of energy conservation applications and alternative energy sources. Among other activities, the Resource Center hosts a variety of educational programs and publishes the *Sunjournal*, a quarterly energy periodical. The annual Aspen Energy Forum is one of the major activities of the Resource Center. The theme for the Aspen Energy Forum 1978 is "Humanistic Choices" and will be held May 26, 27 and 28, 1978.

The Roaring Fork Resource Center expresses sincere appreciation to the following: The RFRC support staff, Sigrid Strecker, Shari Young, Anne Oakes, Rosemary Koch, Gail Grothmann, Kurt Olson; Forum participants; Pitkin County Commissioners; National Economic Research Associates; and John Brandt of the Central Transfer Company of Illinois for the financial support of this document.

>ROARING FORK RESOURCE
>CENTER
>Gregory Franta
>Kenneth Olson
>Heidi Hoffmann
>T. Michael Manchester

Co-Editor: Gregory E. Franta, AIA.

Gregory E. Franta, AIA, is the co-director of the Roaring Fork Resource Center and a solar architect with Sundesigns Architects in Aspen, Colorado. Greg is also the co-director and co-founder of the Colorado Solar Energy Association. He has designed a variety of solar-oriented structures with a "holistic" approach to architecture. Greg has a Master of Architecture degree in Solar Energy Technology from Arizona State University.

Co-Editor: Kenneth R. Olson.

Kenneth R. Olson is the co-director of the Roaring Fork Resource Center in Aspen, Colorado. Ken is primarily a research coordinator and construction technician developing applications for resource conservation in buildings. He has a Bachelor of Science degree in Building Science from Renssalaer Polytechnic Institute in Troy, New York. Ken is presently pursuing a Master of Environmental Planning degree from Arizona State University.

Production Coordinator: Heidi Hoffmann.

Heidi Hoffmann is the co-director of the Roaring Fork Resource Center and the managing editor of the RFRC *Sunjournal* in Aspen, Colorado. Heidi has a Master of Architecture degree and a Bachelor of Arts (majoring in political science and economics with an emphasis on natural resources, policy and administration) from the University of Colorado.

Graphic Coordinator: T. Michael Manchester.

T. Michael Manchester is the co-director of the Roaring Fork Resource Center and solar consultant and building designer with Sundesigns Architects in Aspen, Colorado. Michael has participated in the design of a variety of energy-conserving buildings. Michael has a Bachelor of Architecture degree specializing in Solar Energy Technology from Arizona State University.

CONTENTS

PASSIVE SYSTEMS

CONCEPTS OF SOLAR ARCHITECTURE, ARCHEO-ASTRONOMY AND AUTONOMY Jeffrey Cook	1
"ARCHITECTURE AU NATUREL" David Wright	17
HUDDLING TOGETHER IN MOTHER EARTH Fredric A. Benedict, AIA	29
KEEPING A COOL HEAD AND WARM FEET Keith Haggard	41
PASSIVE SOLAR HEATING OF BUILDINGS J. D. Balcomb	59
PREDICTING THE PERFORMANCE OF PASSIVE SOLAR HEATED BUILDINGS Ed Mazria	89
ENERGY-PROCESSING BUILDING MATERIALS Day Chahroudi	101
SELF-INFLATING MOVABLE INSULATION Ronald Shore and James Gronen	107
NORTHERN WINDOWS AND SOLAR ARCHITECTURE FALLACIES Raymond N. Auger	111
HEATING REQUIREMENTS FOR BUILDINGS William R. Harmon	115

GREENHOUSE

GREENHOUSE CONSTRUCTION WORKSHOPS
"BARN RAISING STYLE" 119
William F. Yanda

ENERGY FLOWS IN THE GREENHOUSE 123
Herbert A. Wade

WET-DIRT STORAGE FOR A SOLAR GREENHOUSE 133
Joseph B. Orr

AN ATTACHED SOLAR-HEATED GREENHOUSE 143
H. E. "Rip" Van Winkle

ACTIVE SYSTEMS

SOLAR COLLECTOR SIZING 149
Herman G. Barkmann

RESIDENTIAL APPLICATIONS OF HYBRID SOLAR THERMAL/
SOLAR PHOTOVOLTAIC ENERGY CONVERSION
TECHNOLOGY 161
Joel DuBow

EUTECTIC SALT AS A SOLAR HEAT STORAGE MEDIUM . . 173
Don M. Harvey

CASE STUDIES

ARCHITECTURE, THE SUN, AND THE ROARING FORK
VALLEY 179
Gregory Franta, AIA

PROJECTS IN TELLURIDE 189
Dean Randle

THE COMMUNITY COLLEGE OF DENVER AND OTHER
SOLAR PROJECTS 201
John Anderson

ALTERNATIVES

BASELINE DESIGN OF COMMERCIAL CENTRAL RECEIVER
SOLAR POWER PLANT 209
Floyd A. Blake

WIND POWER AS A VIABLE ENERGY SOURCE 227
Stan. H. Lowy

WOOD FOR ENERGY 241
Ja Densmore

THE MEANING AND APPLICATION OF GEOTHERMAL
ENERGY 249
  Glenn E. Coury
ENERGY INDEPENDENCE THROUGH METHANE 255
  C. E. Tomson, Jr.
COMPOSTING TOILETS: A VIABLE ALTERNATIVE 271
  David Del Porto

EDUCATION

THE ROARING FORK RESOURCE CENTER 287
  Gail Weinberg
ASPEN COMMUNITY SCHOOL ENVIRONMENTAL
LABORATORY 291
  Paul Rubin and John Katzenberger
WILDWOOD SCHOOL: A PLACE TO EXPLORE THE
ENVIRONMENTAL ETHIC 295
  Bob Lewis
MEANING, ECOLOGY, DESIGN, ETHICS 301
  Wallace W. Wells
THE FARALLONES INSTITUTE—EXPERIMENT IN
APPROPRIATE TECHNOLOGY 307
  Alison Dykstra

INDEX 329

CONCEPTS OF SOLAR ARCHITECTURE, ARCHEOASTRONOMY AND AUTONOMY

Presented and prepared by: Jeffrey Cook

SOLAR BUILDINGS

Why do solar buildings have some magical appeal? We are not particularly interested in a building when it is heated and powered by natural gas, bunker oil #2 or hydroelectricity. These energy forms seem to matter not a twit in conception of a building so long as it works. But, a building based on the sun is blessed. It implies a special conception--the promise of an almost timeless relationship to the cosmos and the potential of an infinite continuum of vital energy. It is a reaction not based on form or expression, but on some abstract idealism. Solar architecture is building conceived in the womb of natural processes. It is design oriented to nature, both Earthly and Heavenly.

NATURE'S DRIVING FORCE

The sun is our closest and most obvious star. Its diurnal appearance defines the rhythm of the days with the warmth and light of its nuclear fire. The motion of the seasons is revealed by the moving point on the horizon of apparent sunrise and sunset. The sun's sequential processionals from dynamic horizons through changing sky domes are human marks of time on earth. For instance, in the language of the Maya, only one word, "kin," means "sun," "day" and "time."

The life force of the sun is basic to the food chains and weather systems that drive the earth's environments. It is a power so obvious, so fundamental and so dependable that mankind has alternately ignored it and worshipped it. Yet, life without sun is inconceivable to earthlings. When scientists discuss life on Mars, it is with the presence of Earth processes as powered by the Sun as parameters. The sun is nature's energizing force on earth.

WORSHIPING THE SUN
Many early cultures worshiped the sun. They rejoiced not just its warmth and light, but also in the precision of its predictably changing path across earthly skies. Early man celebrated the sun not in consumption, but in conception.

Thus, the first architectural expressions of solar reverence were based not on the thermal or mechanical advantages of the sun, but on the spiritual. Throughout the ancient world, early man practiced naked-eye astronomy. He recognized the subtle daily translocation of the sun and sometimes the reflection of the sun in the moon in all the arts, but especially, and most permanently, in astronomical alignments whose conscious precision pervades the layout of buildings and even whole cities and landscapes. The exactitude of these images of respect for the clockwork of the cosmos reveal a scientific prowess highly developed well before writing.

ARCHAEOASTRONOMY
Archaeoastronomy is an interdisciplinary field composed of elements of traditionally established fields of astronomy and archaeology. The intent is to understand the nature of man by studying ancient man's oldest and most widely practiced science, astronomy. The cultural-scientific heritage of ancient and often "primitive" peoples, provides man's first and, perhaps most convincing, examples of solar architecture.

STONEHENGE
The first prehistoric monument to be seriously investigated by

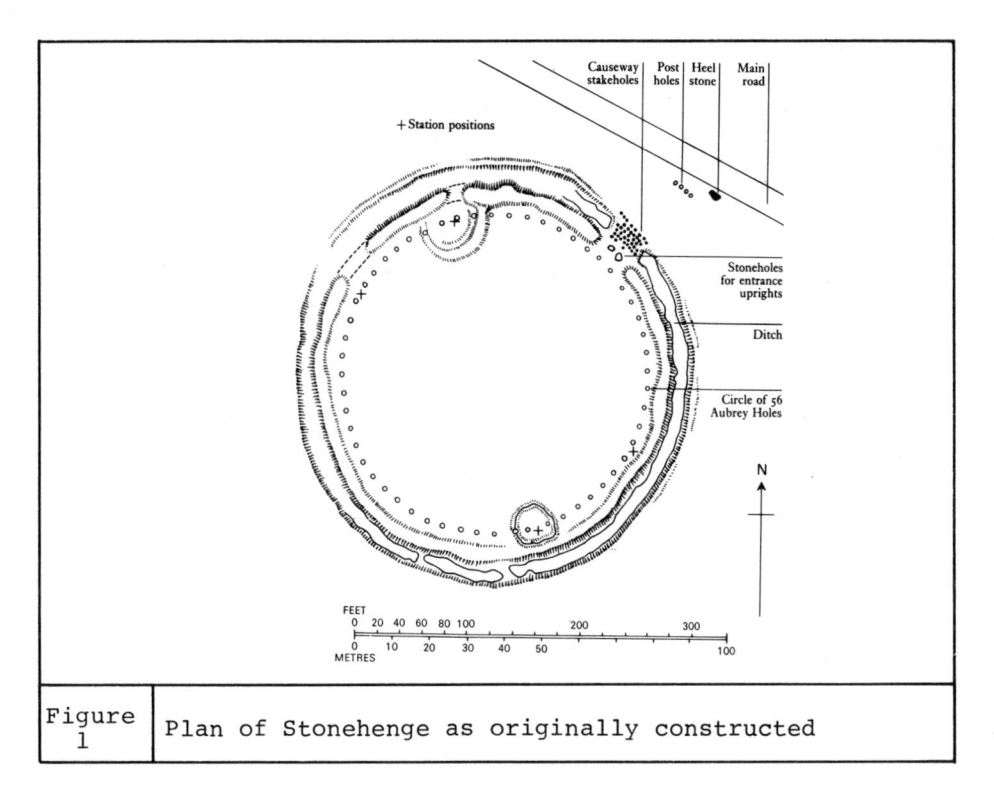

Figure 1 | Plan of Stonehenge as originally constructed

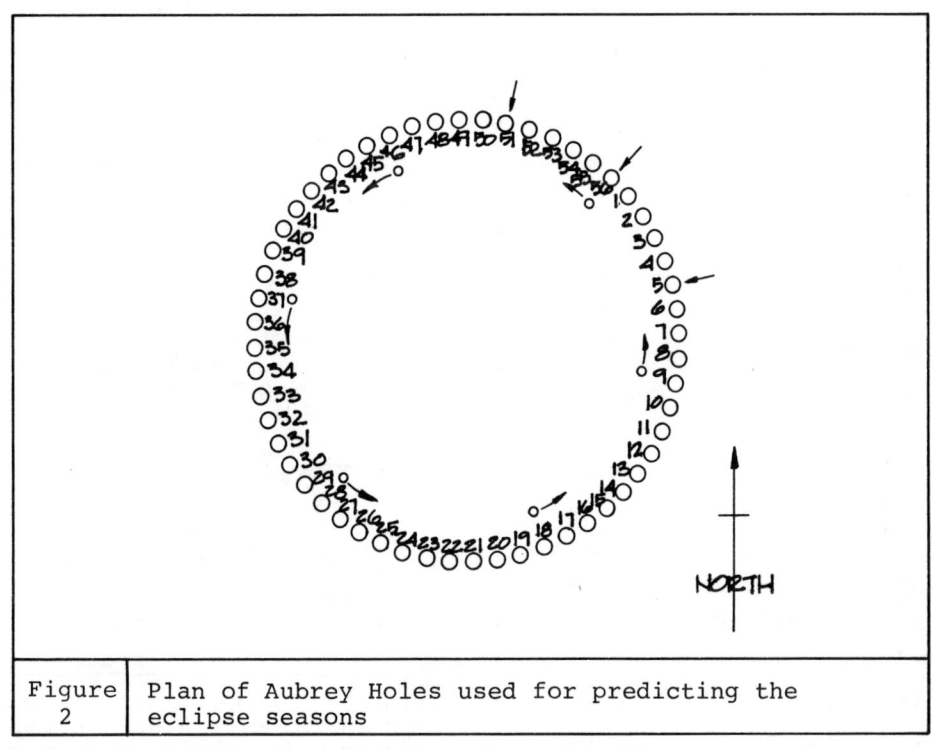

Figure 2 | Plan of Aubrey Holes used for predicting the eclipse seasons

archaeoastronomical methods was Stonehenge, which was first tentatively dated by the exact architectural alignment of sunrise on the longest day of the year--the solstice dawn of midsummer's night.

Stonehenge was initiated in the late Neolithic Age as a solar temple on the isolated and austere downs of the Salisbury Plain in central southern England. It was a barren and undistinguished site, except for its expansive horizons where the successive risings and settings of Sun and Moon could be marked with table-top precision. The several circles of stones, earth mounds and holes that mark Stonehenge today reveal a succession of building periods, each adding a new level of complexity, but with an obvious persistence of function over a period of more than 1,000 0 years.

The first Stonehenge was begun around 2,700 B.C. with the construction of a circular ditch that enclosed an area 300 feet in diameter. Within the resultant bank was a circle of 56 holes. These holes have only recently been understood to be a computer for predicting the moon through its phases, seasons and eclipses. Located along the same circle are the "Four Stations," which mark the corners of an imaginary rectangle, which it is believed was also used for astronomical observations of the sun and moon.

Outside of the circle and its earth embankment was the "heel stone." Although today the heel stone marks the sunrise of the summer solstice, it probably, prehistorically, was a marker before and after the solstice as well as a marker for the midpoint of winter of the full moonrise.

Perhaps 600 years later, around 2100 BC, a new people

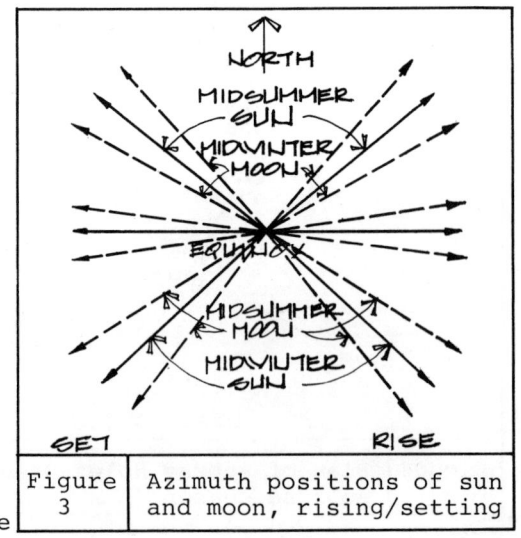

| Figure 3 | Azimuth positions of sun and moon, rising/setting |

Solar Architecture

began an expansion of Stonehenge, especially in the development of an axial approach avenue flanked by earth banks. More important was the trial introduction of a ring of bluestone monoliths brought from the Prescelly Mountains, west of Wales a distance of 130 miles as the crow flies, or by land and sea, a journey of 240 miles. These bluestones weighed up to seven tons and required at least 32 healthy men to move them across flat land on tree trunk rollers. These extraordinary constructions extended and reinforced the original purposes of Stonehenge.

A third construction stage around 2000 BC was the major campaign that left Stonehenge as we know it today. Figure 4 suggests how the center looked when completed. Great hard monoliths of Sarsen weighing up to 50 tons each were brought from near Avebury, 20 miles to the north, and dressed and placed with precision accuracy in a circle with a diameter of 98 feet. Inside the sacred ring a great variety of astronomical alignments were given permanent form. The architectural monumentality of

| Figure 4 | Perspective view of the reconstructed core of Stonehenge in its final stage around 1500 BC |

Stonehenge has long been admired, and it is the best known of all prehistoric ruins.

Predicting obvious astronomical events, such as eclipses and the longest and shortest days of the year have been the basis for priestly disciplines in cultures thoughout time. Seldom have the basis for their mysterious but scientific practices been the basis for an enormous architectural instrument which celebrates the path of the sun and its light on our moon. The astronomical alignments at Stonehenge suggest a complexity of design and understanding that is not usually associated with Stone Age man, especially in a location away from the Mediterranean Basin. The involved geometry of Stonehenge as shown in Figure 5 accommodates the asymmetry of diverse paths in the sky within a symmetrical architectural synthesis.

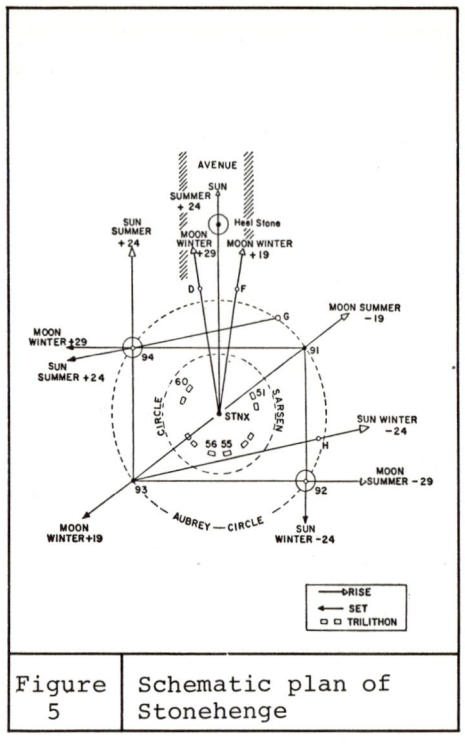

Figure 5 Schematic plan of Stonehenge

ABU SIMBEL

Astronomical studies of the ancient ruins of Egypt have revealed many deliberate solar and lunar alignments. However, perhaps the most memorable is the orientation of the temple of Rameses II at Abu Simbel. Although built after writing had been invented, there is no written record or inscription to confirm the priestly design intentions.

The immense cliff sculptures and rock cut temples at Abu Simbel are considered one of the grandest of Egyptian constructions. Figure 6 a&b give some idea of the scale. The seated figures of Rameses II are 70 feet high. Located on the Nile far to the south of the ancient kingdoms, it was a major monument built in

Solar Architecture

the center of conquered Nubia. The temples at Abu Simbel were moved to higher ground during the 1960's, because of the permanent flooding of the Nile from Lake Nassar by the high Aswan Dam. However, the original alignments have been retained.

| Figure 6 | Ground plan and section of Rock-Temple at Ipsamboul |

Construction was begun around 1270 B.C. by Rameses II well after his accession as Pharaoh of the two kingdoms. The site was selected and the astronomical position of dawn was anticipated, so that on 21 October, 1260 B.C. (10 years later, on the first day of the civil calendar), the "first flash" of sunrise penetrated the axis of the temple to illuminate the figures of three pharaoh god effigies at the end of the sacred sanctuary 180 feet deep inside the living rock. The fourth effigy, the god of darkness, received no light. The date celebrated the 30 year Jubilee of Rameses II's reign and began the process of celestial deification during his lifetime. Today, electric spotlights daily simulate this once-a-year event. Even on 21 October, the sun no longer penetrates--a door has been placed in the temple blocking the light.

An adjacent little chapel dedicated to the sunrise god Ra-Horakhety confirms the ancient Egyptian accuracy in solar prognostication. Its sculptural development of four baboon statues, obelisks and scarabs are all symbols of the sun god. Its separate orientation to the main temple is seen in this. The ascending stairs and raised altar are aligned on the rising sun of the winter solstice--the critical shortest day of the year when the sun must turn around to return.

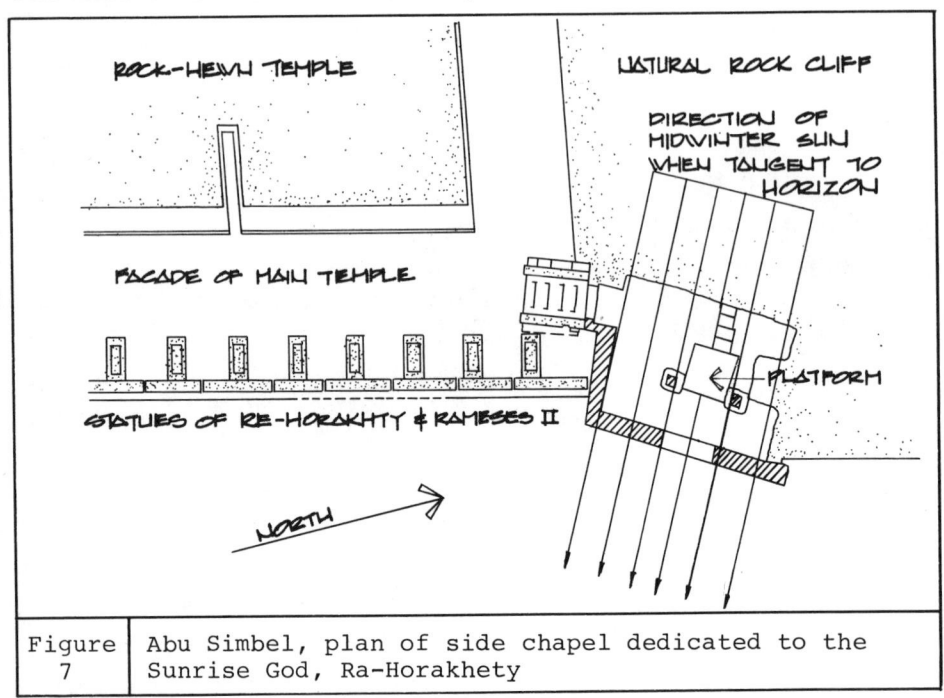

Figure 7. Abu Simbel, plan of side chapel dedicated to the Sunrise God, Ra-Horakhety

CHICHEN ITZA

In the Americas, the sightings through architectural openings of the so-called Caracol (Snail), at Chichen Itza, has encouraged the realization that this handsome spiral Mayan structure is an Observatory. Like Stonehenge, it was also renovated and rebuilt successively after use. Its present form was adjusted by the invading Aztecs around 1000 A.D. For instance, Window 1 as shown in Figure 8, aligns with the equinox sunsets. In addition, the maximum northerly moon set and maximum northerly Venus set also fit within the sights of the broad slot.

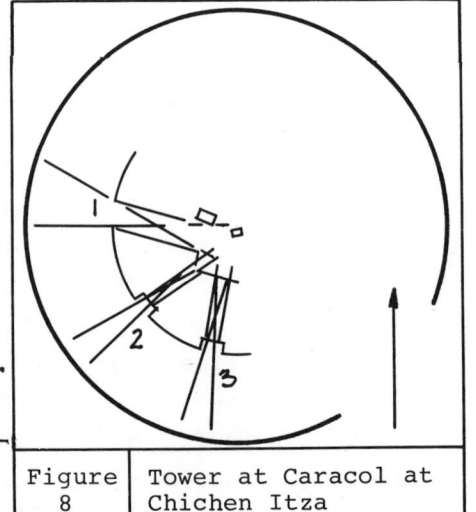

Figure 8. Tower at Caracol at Chichen Itza

Solar Architecture   9

As elsewhere in Mayan ruins, the window alignments of sunrises and sunsets are reinforced by other astronomical phenomena. Timed patterns of sun and shadow are achieved architecturally, only to be revealed on celebratory days. These solar definitions of the order of gods in the universe are complimented by alignments with other more subtle stars and constellations.

SOUTHWEST UNITED STATES
Astronomical alignments are characteristic of a religious and symbolic recognition of the sun throughout the world. The layout of the pueblos of the southwestern part of the United States are rich in such evidence. Here, the position of Sunwatcher continues to be an important responsibility in the organization of present day pueblos. Prehistoric petroglyphs of sun symbols are more frequent than any other subject.

Perhaps the most remarkable solar structure of the New World is Pueblo Bonito in the Chaco Canyon of New Mexico. As seen in Figure 9, the summer solstice dawn defined the ends of the

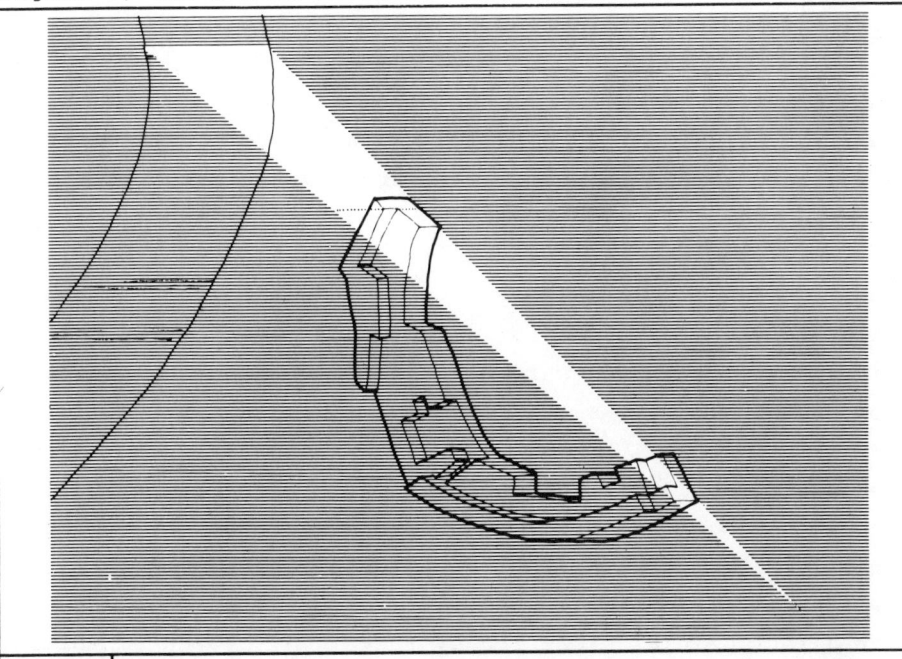

| Figure 9 | Old Pueblo Bonito - the first light of the summer solstice |

Pueblo. Not only are the alignments of the solstice reflected in the form of the megastructured community, but recent thermal studies have confirmed that the cup shape of the pueblo form is an effective solar collector. Astronomy and comfort are simultaneously synthesized architecturally. The refined shape of the structure provides a natural means of increasing thermal gain in winter when it is most needed. Again, its solar performance, both thermal and astronomical, was perfected by successive additions to the initial building with a construction period that lasted over 150 years, from approximately 919 A.D. to 1067 A.D.

These architectural links of early man between the shape of buildings and the earth's most important star are antecedents of 20th century attempts to recognize the sun. But, in our time, it is the thermal and mechanical advantage, not the confirmation of the mysteries of the universe, which are dominant.

Celestial order has been replaced by economical ordering. For us, it is physics, not metaphysics.

SOLAR BUILDINGS
The first buildings to be called "solar" were houses generated in the late 1930's in the Chicago area by the architects George and Fred Keck. They were not involved with hardware or equipment, but with the thermal results of architectural configuration. In simple terms, they placed windows to the south with calculated overhangs to provide seasonal direct solar heat gain. Heavy draperies were pulled at night to keep the heat from escaping.

The 1940's and 1950's were a time period when stylistically the glass box became the architectural order of the day. Experimentally, the most important architectural research carried out during that period was in the field of solar shading and selective direct gain as pioneered by the Olgyay brothers. Simultaneously, the definitive scientific works which established the calculable parameters of collecting the sun's heat were derived and published during the 1950's. In parallel, a small number of interesting houses built primarily in the southwest demonstrated the application of these principles.

Active and Passive

To us, today, the term "solar architecture" implies buildings whose design integrates the thermal, directional and seasonal aspects of the sun. When the latest fascination with applied solar energy was revived at the beginning of the 1970's, there emerged a double approach. On the one hand, direct gain and integral thermal storage in designs by David Wright for his own house and for Karen Terry at Santa Fe demonstrated an architectural approach which is now called "passive." As shown in Figure 10, the design uses solar windows and built-in thermal storage to maintain a satisfactory interior environment. Alternatively, "active" systems utilizing complex hardware, specialized equipment and a high degree of technical control were early demonstrated by such houses as Solar I of the University of

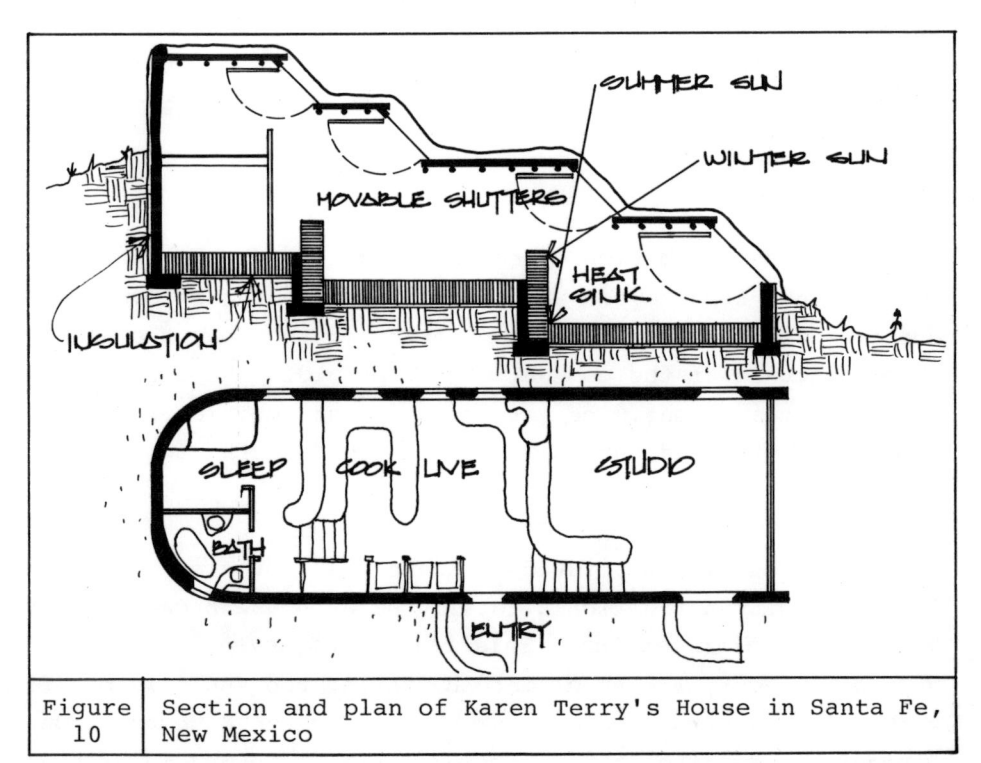

| Figure 10 | Section and plan of Karen Terry's House in Santa Fe, New Mexico |

Delaware and the Decade 80 house of the Copper Development Association built at Tucson. To varying degrees, all of these houses were dependent upon the sun for domestic space heating. In those houses which were owner-built and self-financed, the backup systems were minimum. Thus, the percentage of thermal comfort delivered by the sun had to be relatively high. Alternatively, both Decade 80 and Solar I included full back-up systems which added to their already substantial capital cost.

Autonomy
More ambitiously, both Decade 80 and Solar I attempted not only thermal comfort, but also limited on-site electrical production. Thus, the building and its equipment aspired to become an integrated total-energy machine. Conceptually, once such an investment has been made, a building could be energized autonomously, independent of local supplies.

Against the inherent frugality of home-made and traditionally based passive solar designs which attempted personal independence and self-sufficiency, these highly technical solar solutions have been accused of mistaking indulgent consumption for higher standards of living.

BEYOND AUTONOMY
The most recent ambition of solar buildings, whether active or passive, has been beyond autonomy. By designing an overly productive building, the surplus might begin to repay the energy superstructure of the community that provided the original capital. At the moment, the largest and most ambitiously built examples of "beyond autonomy" can be found north of the border, in Canada.

Provident House. The Provident House near Toronto, completed in 1976, is by definition a totally solar house because it provides 100% solar heating. This feat is achieved in a cold, grey and lengthy winter climate by saving the sun's warmth from summer in a generous water tank that fills the basement. On an exposed, windy site, the excavated earth is formed in grass-covered berms that screen the wind to provide a shelterd, modified microclimate. A greenhouse supplies a tempered, complimentary social space, as well as food growing area. Prudent design provides a spacious, light interior within a highly efficient enclosure that has a minimum of window openings.

P.E.I. Ark. A more ambitiously solar productive building is the New Ark on P.E.I. in the Atlantic Provinces of Canada, completed in 1976. Accomodations for an extended family are integrated with a solar greenhouse that permits intensive production of both food fibers, as well as fish protein. The complex planning is revealed in cross sections of the building as seen in Figures 11 & 12. The growing season is extended year around. It is an integrated food factory in which lifestyle and life-support are merged. Intense capital investment provides an experimental ecological bioshelter. An adjacent windmill array supplies electricity, with the surplus fed into the power grid.

REGIONAL SOLAR UTILITIES
Beyond these surplus solar support systems, the next logical step is toward community and regional solar support and exchange. The new regional bus facilities (RTD) in Denver, Colorado, are an example on such a scale. It is an economical scheme, both in construction and in use. Earth-embraced walls provide an economical enclosure with thermal resilience. An array of tilted solar collectors crown the roof. A comfortable garage and maintenance space is provided on a vast scale with a minimum of fossil combustion either on or off the site.

Similarly, the new Community College of Denver (North Campus), now located in open countryside northwest of the city, also is becoming a regional facility, not only educationally, but also as a community landmark. It is a public utility in several senses.

LANDMARKS
All of these examples, both prehistoric and contemporary, are not isolated ideas. They are landmarks in an emerging new

Figure 11 | Section through P.E.I. Ark, a year around solar greenhouse, aquaculture farm

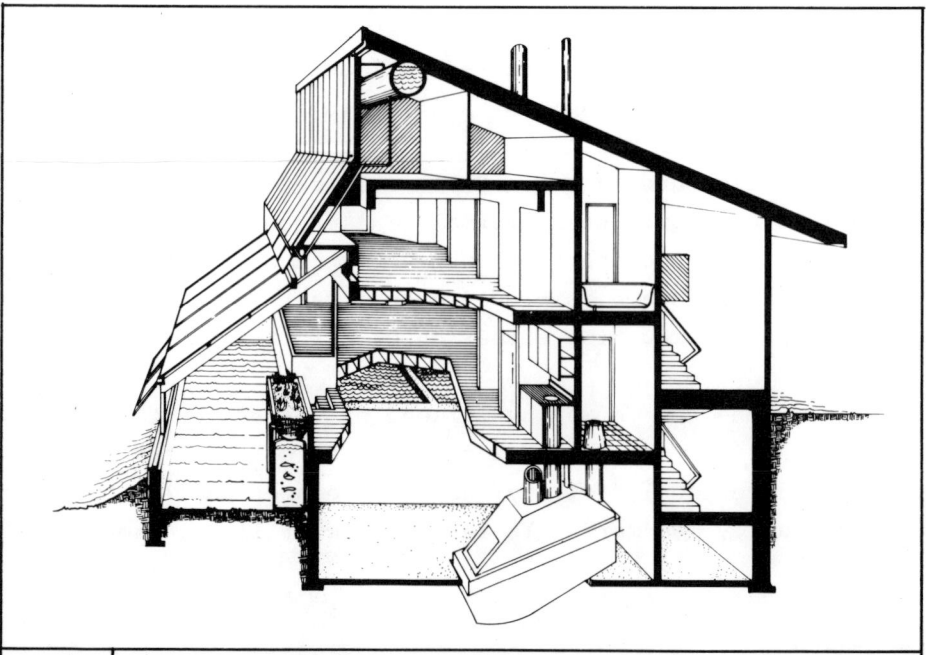

Figure 12 | Section through P.E.I. Ark - integration of human, solar, biological and thermal systems

design discipline of buildings. They mark important ethical reorientations of expanding influence that go well beyond man's physical need to keep his body warm. They re-establish, architecturally, man's inescapable dependence on the ever-present sun as central to life processes on earth.

BIOGRAPHICAL SKETCH

Jeffrey Cook is an architect specializing in energy-conscious design. He is a professor of architecture at the College of Architecture, Arizona State University, in Tempe, Arizona. Mr. Cook is also a member of the Board of Advisors of the Roaring Fork Resource Center in Aspen.

"ARCHITECTURE AU NATUREL"

Presented by: David Wright
Prepared from a transcription

I would like to speak about how Mother Nature would approach an architectural design commission, were she to be granted one by a human being. I would also like to look at the intricacies that are involved in designing for different microclimates, optimizing all that nature gives us as tools.

I recently had the opportunity to look at a new project in Southeastern Oklahoma. It is the first time I have ever been there; it is a beautiful place and probably the most challenging place that I have had the opportunity to attempt a passive solar design. I will review some of the weather conditions for this particular site. The tornado winds may reach velocities of up to 140 mph. It rains quite often. Therefore, another parameter to deal with is flooding. The temperature can go from 20 F to 100 F in one eight-hour period and up to 85% humidity and 110 F may occur.

In Oklahoma, heating is not really the main problem. Although there are blizzards, snowfall and temperatures down to -20 F at certain times of the year, the main problem is cooling. Design conditions are approximately 3000 degree days heating and 5000 degree days cooling. We are going to try to cool and dehumidify passively. That is how nature would do it.

One of the things that was most interesting to me was that right on this particular site, a badger was living. The fellow that was with me said, "This is a badger hole and don't stick your hand down it because he is liable to bite it off and come out fighting." Despite his bad temper, the badger had adapted quite well to this particular environment. He is one of the larger mammals in this area. There aren't many deer or antelope. He very wisely buried his home into the bank of the creek, and that keeps him cool in the summertime. There were trees shrouded around the creek causing a definitely cooler microclimate in that area. He would also be warmer in the winter by virtue of burrowing into the earth's crust. The badger gave me some ideas. One approach we might use is digging into the earth, because we don't have a high water table and dirt is free.

When looking at things this way and putting yourself into this attitude using Mother Nature's glasses, you see the natural technology of things that tend to balance out the microclimate. I am not advocating abandoning technology as we know it, even high technology. There are some absolutely vital products and ideas that we have developed that will help to sustain our lifestyles and the quality of life for a long time. Materials like glass, insulation and a lot of simple elegant devices which are products of man's technology, help us design and build a passive solar house. The natural technology and materials have been overlooked, however, to a great extent. For instance, take adobe: it looks nice, doesn't use much energy, holds buildings up and makes a good heat sink. Adobe symbolizes the potential of many other materials that we have around and could use more beneficially.

One of the approaches to this natural process of design is a system much like the one Ian McHarg has developed: designing with landscape patterns where they occur. By analyzing the environmental factors (i.e. systematic overlays of soil types, noise, pollution factors, drainage patterns and many other things) where man might have to build a freeway or new city, the patterns will give you a picture of where that highway might be built or new town developed. In the past, particularly here in America, we had the old pioneer spirit. The person who found the first well, built a building, and someone else built next to it. That became the main street, and the town just kind of went on from there and was gridded in rectangles. There wasn't much logic other than real estate values applied to how the land was used. Now we are saddled with that. That is why our cities are so difficult to retrieve from their decline, and why many small towns require a lot of energy to get from one point to another. It was just historical expedience on our part.

Now we have come to a time where we can take advantage of our technology and knowledge of the natural factors and put them together to get the best of two worlds. One way of doing this in architecture is what I call "microclimate design."

(Editor's note: Color transparency slides aided the following presentation.)

A wasp's nest is a really good example of insects using organic materials to create a very complicated structure with very low energy in terms of material use and high concentration of energy in terms of manual labor. It is a great form and is perfectly in tune with its environment. If you look at that and observe how a great many insects are living in one structure in a very organized way, you would see a vast difference in structure and organization compared to one of our towns. When you think about the chaos that we have got, you might think that the wasps have something going for them!

Definition of a particular microclimate shows that it is unique to others. Each specific piece of land has some kind of weather pattern that is unique to that area. The task is not just a matter of putting a solar house there with a bunch of solar collectors on it and insulating it. You must study all the factors that are working on that particular site and understand them. If you understand them well enough, you can design a structure with less material that will last longer and will perform better. The landscape and climate dictate the rules. It is pretty hard at first to understand this idea. We are used to buying two-by-fours, going out, pouring a concrete slab and putting up a house. But now we are getting a little more "hip." We are putting less windows on the north side, solar collectors on the south side and even air lock entries to achieve better performance. But we have a long way to go. Analyzing the landscape and the microclimate provide the direction.

Upon assessing a given site, first look at man's influence on the land. How much damage has man done? How does the proposed building affect density and pollution problems that man has created. What can you do with it? You have to look at urban, suburban, rural and natural state. You should make a value judgment as to whether it should be used at all. Many of these natural places should be left as is.

The amazing forces of nature which affect the northern California coastline are far different from those of Saudi Arabia, for example. One of the things that a site in Saudi Arabia might have is a tremendous amount of raw material in the form of sand that can be used for building structures and an awful lot of sunshine. My attitude is that you can't have too much heat. Heat is a fuel. It is valuable. Sometimes it might be uncomfortable, but you can handle it. You are lucky to have enough in there. It is easier sometimes to get cool than it is to get warm.

Latitude is a landscape characteristic which defines your relationship to the equator. It tells you something about how a building or a structure ought to be shaped in order to take best advantage of the sun. At the equator you don't need as large a collector surface. It would be almost horizontal for optimal year around incidence, whereas the farther north or south you get from the equator, the greater the optimum collector angle would be. When you get up to Canada the best orientation is a vertical collector surface. Latitude can tell you about shapes of the buildings and how they fit into the landscape. Each building will react to it.

Solar Architecture

Architecture can be a lot more responsive than it has been in the past! It is technologically terrific that a heat pump can be plugged into a building, as is traditionally done, and take care of the heating or cooling load. Right now the heating and cooling bills are averaging about $100 a month in this area of Oklahoma. One of the local utility people said that utility rates are going to triple in the next three to five years. Then utilities will be rationed. The electric meter will be set, so a household will have a lifeline limit of electric power. When you pass this ration, the meter goes off and you have lost your lifeline. Heating is the most obvious aspect of solar energy, but solar cooling is also a possibility. Some designers are specializing in active systems, some with passive and some with hybrid systems. This is just the tip of the iceberg, however. Solar energy is basically a conservation effort and that hasn't struck home to all of us yet. One thinks, "oh yes, we are going to cut out electricity and stop using the natural gas." But there are many other aspects which make solar energy attractive to us. For lack of a larger vocabulary, let us call it "sex appeal." It is interesting, adventuresome and is possibly a new frontier. You can invent new things. It becomes technical, architectural, "first kid on the block to have a solar house" kind of attitude. There is a lot more to be done than that. With this particular house in Oklahoma we will attempt to cool and dehumidify during the summer, heat during the winter, supply domestic hot water and some electricity. We don't know how much electricity we can generate, but we know Oklahoma is a good "wind state," if your wind generator doesn't disappear during a tornado.

When approaching a design in a particular microclimate, look at it from the point of view of Mother Nature. Try to step back and not be the kind of human technician who is conditioned to certain responses and educational factors and has a limited IQ. Just sit back and be as objective as possible. It is very difficult, as we all have our human limitations, but it is fun to try it.

Pueblo Bonito was a very ingenious development that developed over a period of time. The people varied the materials and the placement of their heat sinks until they found out what worked best. They weren't in a big rush to solve the problem in two months, so that they could get a building permit before the next winter.

Upon noticing a cat sleeping in a violin case one day, it occurred to me that form doesn't necessarily have to follow function. And that can be true in buildings also. If you design a building for a particular microclimate, perhaps there is an optimum form that should be used for heating, cooling or dehumidifying, whatever the consideration. Many times, human beings can adapt to that form. The igloo is an example of that. It functions well using low-energy materials in a very harsh climate, in a very sensitive, expedient way. The people live relatively comfortably, not by our standards but by their own. They would probably suffocate in a tract home in Denver, Colorado or San Jose, California.

Just about every place you are, there is a certain amount of indigenous building material. In a few years, if not already, using earth and local materials will be the most economical way to build structures. It's nice that we have tremendously strong steels and exotic metals to build highrise buildings and skyscrapers. But look at a normal tract house. There are an awful lot of them made out of very expensive materials, in terms of energy.

To give an example, last week we calculated the amount of energy in kilowatt-hours it takes to make an 8x8x16 concrete block versus an adobe or rammed earth-block of the same volume. The ratio in kilowatt-hours is 334 to 1. The concrete is extremely energy wasteful because the material has been processed, transported and gone through a lot of changes before it arrives at the building site. The rammed earth is probably already on the site and requires little modification. Therefore, it is going to be more labor intensive than industry intensive. Granted, there

Solar Architecture 23

are certain tradeoffs that you have to look at: How long is it going to last? What is its bearing strength? What is its character and charm; it's value in terms of aesthetics? I don't see how one can justify concrete block over the potential of earth for small buildings. Even if the earth wall is required to be five-times thicker than a concrete wall, a lot of energy still has been saved. There are many other materials that can be used (i.e. stone, sand, sandbags, fired clay and sundried clay). Adobe and clay are probably the most widely used building materials in the world. Only in the most technically advanced countries is it ignored.

Structural shape, suited to the terrain, is another landscape characteristic. If situated on the north side of a hill, you can build your structure so as to capture even the small amount of solar energy available and make it work. The profile of the land is going to somewhat dictate the profile of the building just as latitude did.

We are talking about microclimate design and how to best design a solar house. Inevitably, view is going to be a part of your design decision; whether to build on that piece of land and how to orient it. I have seen solar houses with direct-gain systems on the south and large view windows on the north. This is self-defeating. You have to decide when you are going to compromise. We don't always have to have a good view. We can have private views and special-made views. Even if you are living in a place where you think the environment has declined to a point where everything turns you off, you can still survive there with an interior rock garden or other private aesthetic. View is important too when thinking of building on the landscape. What you build is going to be there for a long time. Therefore, build as beautifully as you can; plant some trees and help regenerate it.

Vegetation is very important and is often overlooked. You might want to block the western sun in the summertime with a deciduous tree. Vegetation can change the microclimate around a structure from all sides. Evergreen trees may block strong winter winds from the north and help insulate the building as a result.

Soil conditions are tremendously important, and most of us don't even think about it. We go out, get a building permit, and the inspector says, "you need footings so big, so the house won't settle." However, there is really a lot of potential for avoiding problems by analyzing the soil. Whether the soil has water problems or is good to build with may be determined. If it is going to get wet and expand upwards, it could break your walls apart. Maybe you can use it as a heat sink or as a cooling medium. A study of land types can indicate the type of architecture that is best suited for any place. It is not always advantageous to go underground. In the tropics, for example, you must consider humidity and moisture. In the desert, it mades tremendous sense to go underground for stabilization of temperature.

Other considerations of the microclimate should also be analyzed:
1. The water shortage is the next crisis. One should know where the water is coming from.
2. Temperature can vary from one side of a mountain to the other. Precipitation affects your land site and where it goes when it rains should be studied.
3. Humidity is also an important factor in providing comfort. You are not as comfortable at 65 F with 50% as 60 F with 10% humidity.
4. Air motion should be observed as to how it interacts with the land. Even with very low air motion, we can design structures to take advantage of a breeze or cooling by solar induction.
5. Patterns of weather cycles should be looked at as to how they work and frequency of cycles.
6. Shielding and catching sunlight and using it to best advantage should be considered.
7. Pollution is something we should all know about in order to combat it.

Solar Architecture

8. Earthquakes, floods, fires, hurricanes and tornadoes, etc. are acts of God. Don't build in the flood plain and consider what will minimize effects from other disasters.
9. Look at man's structures in the particular region. Analyze early buildings and traditional structures and use the wisdom of previous inhabitants.

I tried to protect my house, "Sundown" at Sea Ranch, California, from the wind and provide for some outside area as well. I also tried to optimize the solar heating and cooling capacity.
If the house starts to overheat in the fall, our first line of defense is to open some vents at the top of the windows. That causes quite a bit of convection, and we get air motion which removes the heat. We can open one of the skylights and pull cool air through the house. Usually we can bring the temperature from 85 F down to 70 F in a half-hour. If we can't control it that easily by convection, then we can pull shades down to stop the solar energy from getting in. We open vents at the bottom and convectively pull cool, fresh air into the house. The solar curtain is three layers of aluminized polyester sandwiched between bright-colored cotton material. Our only auxiliary heat is a wood-burning stove that we used only two nights last winter. The house went from a daily high of about 75-80 F, depending upon how hot we let it get before we started venting it off, down to 65 F at night.

Last winter we experienced deep-space radiation due to the cloudless skies (little rain), and it was colder than normal. This meant that more sun heat was needed during the day. Nature has a way of self-balancing. The heat sink is the back wall which is poured concrete and the floor which is brick on sand. With shallow building depths, I put insulation under the floor and use it as a moisture membrane as well as isolating the heat sink to the inside, particularly as the climate gets colder. Although our house is underground, it is very light and airy due to the solar windows. A view from the West shows a solar facade with an overhang. The overhang is where the vents dump air to the

Figure 1. Southeast view of "Sundown" at Sea Ranch, California; designed by David Wright

Figure 2. North elevation of David Wright house at Sea Ranch, California

outside. Hot air stratifies and escapes out into the eave. Along the bottom, on all the solar gain windows, we have awning windows that open out to give us air intake for cooling. These windows are actually flat-plate, thermosyphon collectors which furnish domestic hot water.

My house also has a sod roof which helps to minimize heat loss. A sod roof should be sprinkled, as there is not as much depth of soil.

Another house we built on a different site, has a much different shape of structure and material use. The two microclimates are totally different. We used the materials at hand and took advantage of all climatic factors to create architecture that in each case interacted with the microclimate. When this approach is fully understood and applied, new kinds of buildings will be created that compliment nature and make man's survival easier and more logical.

BIOGRAPHICAL SKETCH
David Wright is an environmental architect from Sea Ranch, California and is noted for his outstanding design of passive solar systems.

possibly fatal environments, usually either very hot or very cold ones. The presumption is that something simply has to be done by man in these environments and that they cannot be cost-effectively changed too much. They must just let him do what he does without killing him. Here is a sample experiment in the field from the 1967 ASHRAE Handbook of Fundamentals:

> Continuous exposure of nude men to an ambient temperature of 8 F for 3 to 9 days caused a two-fold increase in metabolic rate, a significant increase in pulse rate and marked loss of body protein which persisted several days after the subject had returned to normal room temperature. Under these conditions, food intake is most important. A reduction in calories from 3000 to 1500 per day caused a marked reduction in physical work capacity, as judged by an all-out treadmill test.

Cold is apparently more tolerable than heat _in extremis_, since "long-continued" exposure to high temperatures rather than low ones "is known to lead to deterioration of performance and morale even in the most fit and well acclimatized individuals." The safe exposure of working men to temperatures over 140 F at 30% relative humidity, for example, lasts less than 1/2 hour. This is a long way from the _Kama Sutra_, but it helps to tell us what we can do if we have to.

Optimization research uses enhanced task performance, rather than survey responses or life and death as its measure. Conditions are more "humdrum" in optimization research, which usually takes place in offices, schools and factories. The optimizer's technique is to vary one factor at a time in the environment and to measure any resulting increase or decrease in units of work performed. Noise level, lighting level, temperature and air change have all been studied with various results, tending mainly to the conclusion that the effect of the environment on what you are doing depends on what you are doing. In early tests, it appeared that _any_ change in the environment led to improved performance. The subjects were responding to change itself and to the attention paid them by the experimenters, who cared enough to experiment. This hump at the start of the experiment can now be ironed out of the results and reasonably long term change identified.

If you are interested in optimization research, you will enjoy the work of Richard Stein FAIA, New York. Stein began with a concern for energy use in school buildings and has arrived at some interesting conclusions regarding lighting levels--they are too high--and maintenance--it is crucial. This last theme has lately been taken up by the National Bureau of Standards, who foresee a new, improved and streamlined man/machine interaction as the key to buildings that save energy in fact, as well as on paper. They have an interesting table to offer enumerating the things that we do better and that they (the machines) do better. Among other things, they are better at routine work, and we react better to extremely subtle cues and highly improbable events. With luck and adequate further research, we may eventually be able to _prove_ that we are human.

Daniel P. Wyon has several intriguing things to say. First, his studies indicate that, given a choice, people (at least, people in Sweden) will opt for fresh air or at least different air at regular intervals over the day, even though (or perhaps because) the new air is cold and depresses the room air temperature for a few moments. This was established in two different ways, once on purpose and once by accident. In the accidental case, the actual energy use of an office building seemed so different, and so much higher, than its predicted energy use, that the difference was investigated. It turned out that the office people were throwing the windows open every half hour or so throughout the day in order to get some fresh air. Morale, infiltration and energy use all went up together, leaving the workers happy, healthy and expensive. In the controlled case, office occupants had access to a button which would trigger a similar drastic air change as often as it was pressed. It was pressed with the same sort of regularity.

Wyon's next interesting observations are that both more numerous "spontaneous expressions of comfort" (whatever they may be) and faster task performance occur when the room temperature swings a bit. Although the experimental data are few, the indications are that a temperature swing of about 4 F every 9 minutes or so

Solar Architecture

basis. This technique is used to characterize the predicted performance of various passive solar heating concepts which use solar gain through windows and thermal storage mass walls. The influence of design is studied by successive computations made with different parameters.

BUILDING MODEL

The type of thermal network model utilized for the analysis is shown in Figure 11. This represents a building with a concrete thermal storage wall located behind vertical double glazings. Node 5 represents the temperature of the air between the inner glazing and the wall. Nodes 6 and 11 represent the wall outer and inner surface temperature. The mass of the wall is equally divided between Nodes 7, 8, 9 and 10, which represent the temperatures within the wall at distances of 1/8, 3/8, 5/8 and 7/8 of the wall thickness from the outer surface. Node 2 represents the room (globe) temperature and Node 1 represents the outside ambient air temperature.

Solar radiation on a horizontal surface, ambient air temperature and wind velocity are the input variables to analysis, given at hourly intervals. Solar radiation on a vertical surface is calculated by separating the horizontal surface data into direct and diffuse components and applying the proper geometrical transformation as described by Liu and Jordan.[8] A ground reflectance of 0.3 is assumed. The separation of solar radiation into components is done using the technique described by Boes.[9] The transmittance of the glass is calculated as a function of incidence angle using the Fresnel relations and accounting for glass absorptance. The transmittance at normal incidence is 86% per glass layer. The radiation transmitted through the glass is all absorbed at Node 6.

At each hour, node temperatures are calculated as required to achieve an energy balance at the node. Energy storage occurs at only Nodes 7, 8, 9 and 10.

The thermal energy flow between nodes is calculated based on the temperature difference and the appropriate U-value. The values of U_2, U_4 and U_5 include non-linear radiation terms. All emittances are set at 0.8. The value of U_2 includes a wind-velocity-dependent convective term. The values of U_3, U_6 and U_7 are a constant 1.0 Btu/hr/F/ft^2. The values of U_8, U_9, U_{10}, U_{11} and U_{12} represent thermal conduction through the concrete.

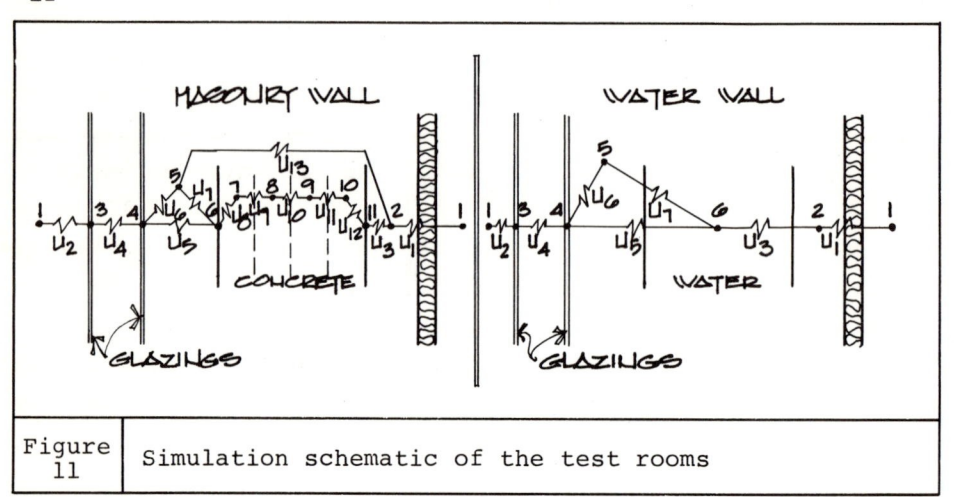

| Figure 11 | Simulation schematic of the test rooms |

In the original version of the masonry thermal-storage wall, as developed by Felix Trombe and his colleagues, vent holes are left at the bottom and top of the wall to allow a natural convection air flow from the room floor level up through the space between the inner glass and the wall and return to the room at the ceiling level. This "thermocirculation" provides a mechanism for instantaneous flow of heat into the room during the day. The conductance U_{13} represents this energy flow path. The volumetric air flow is determined from the following relationship:

$$C_d A_v' \sqrt{g\beta (T_5 - T_2)|/H}$$

where C_d vent discharge coefficient = 0.8
A_d' = vent opening area per unit width,
= 0.074 ft^2/ft
H = wall height = 8 ft
$\beta = 1/T_5$

Various options are possible as follows:

Angeles, CA, July 28 to August 1, 1975 (unpublished). To appear shortly in Solar Energy (Journal of the International Solar Energy Society). Also see: J.E. Perry, Jr., "The Wallassey School," published in the Proceedings of the Workshop and Conference on Passive Solar Heating and Cooling, Albuquerque, NM, May, 1976.
2. P. VanDresser, "Energy Conserving Folk Architecture in Rural New Mexico," ASC/AIA Forum 75 on Solar Architecture, Arizona State University, 1975, pp 12.1-42.
3. F. Trombe, Maisons Solaires, Techniques de l'Ingenieur (3), 1974, C777.
4. F. Trombe, J.F. Roberts, M. Cabanat and B. Sesolis, "Some Performance Characteristics of the CNRS Solar House Collectors," published in the Proceedings of the Passive Solar Heating and Cooling Workshop and Conference, Albuquerque, NM, May, 1976.
5. R. Fisher and W.F. Yanda, "Solar Greenhouses," John Muir Press, Santa Fe, NM (1976).
6. "Research Evaluation of a System of Natural Air Conditioning," Cal. Poly. State Univ., January, 1975.
7. P. Davis, "To Air is Human," Proceedings of the Workshop and Conference on Passive Solar Heating and Cooling, Albuquerque, NM, May 1976.
8. B.Y.H. Liu and R.C. Jordan, "Availability of Solar Energy for Flat-Plate Solar Heat Collectors," ASHRAE, Low Temperature Engineering Application of Solar Energy, 1967.
9. E.C. Boes, "Estimating the Direct Component of Solar Radiation," paper presented at ISES Congress, Los Angeles, CA, July, 1975.

SELF-INFLATING MOVABLE INSULATION

Presented and prepared by: Ronald Shore, James Gronen

REASONS FOR UNDERTAKING WORK

A passive solar greenhouse has been constructed at the Colorado Rocky Mountain School in the lower Roaring Fork Valley. There was a need for a movable insulation that could be easily automated, in order to keep nighttime temperatures high enough for the plants to live. A multi-layered curtain that was constructed of reflective materials seemed to be a good solution for that particular greenhouse. It became apparent that the curtain, if successful, would have many practical applications. Applications for vertical south wall passive systems were further investigated and the unique self-inflation facet of the system was realized. The unique ability of a series of layers having a high reflectivity and low emissivity to separate from one another due to radiant energy is the principle of operation of this curtain. Radiant energy from thermal mass (winter mode) or direct solar gain (summer mode) heats air which rises in the curtain and "blows it up".

Curtains 24 feet long by 16 feet high have been constructed (380 square feet of movable insulation). Storage requirement of the curtain is a space measuring 7 3/4 inches in diameter by 24 feet long. Another unique feature is the ability to cover large areas with effective insulation while keeping the linear feet of crackage (infiltration leakage) at a bare minimum.

This is the first disclosure of the system to the public (patents are being applied for both in the U.S. and abroad). In order to judge the success of the curtain, the insulation values needed to be determined.

PRESENT WORK AND TESTING

The testing of the curtain has not, by any means, been completed. It is still in the experimental stages and will be for some time. However, enough data has been collected in order for us to make some safe conclusions. In determining the insulation values of the curtain, comparative values were stressed, rather than the R-values. This was done because R-values have been misused so much that they have become confusing. In order to get a comparison, two identical boxes were built. They were constructed of 2" thick Dow SM foam and 3/8" plywood. Each box had two sides that were open. On these openings, glass was placed along with the material to be tested. When the tests were run, a 5-gallon drum of hot water was placed within each box. The rate of temperature decline was charted over a certain time period, but more important, at the end of each test the water temperature of each can was checked to see which was hottest. The air temperatures of the hotter boxes were consistently higher, while the glass remained consistently cooler.

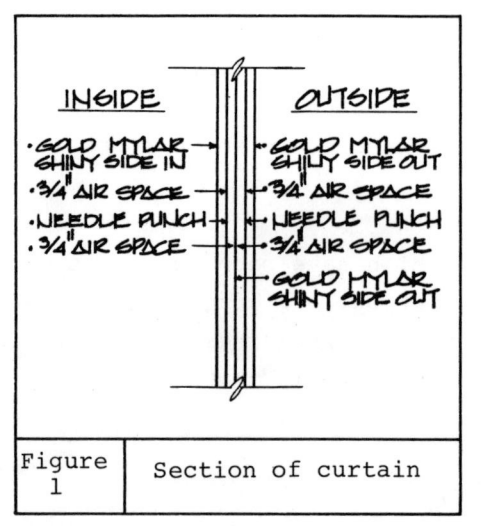

Figure 1. Section of curtain

In each test, some form of the curtain was raced against a material with a known R-value. By testing this way we could say that one material was better than another. In each test, a five-layer curtain was used.

Solar Architecture

with mostly cloudy days, it may be reasonable to glaze all but the north wall or you may choose to keep light levels up during dreary days by the use of fluorescent lights strategically placed (another trade-off situation). Naturally, you have to consider the plants and your use pattern as well. If rapid growth and lots of food production are expected, light may be a limiting factor particularly during the short days of winter. To get the desired production, more glazing or the addition of artificial lights is called for--but keep in mind that more glazing may not give more light! Most light problems occur in the winter since the days are much shorter. The sun never leaves the south part of the sky in our latitudes and as a result little light can be gained on clear days by adding glazing unless properly located.

Thermal storage is the key to maintaining an energy balance in the greenhouse. It needs to be noted that the mere presence of large masses of heat-storage material in the confines of the greenhouse does not indicate that the heat is being stored. After all, the floor and the earth under it form a large storage mass. Its contribution is not very great. But it is of some benefit, if the foundation is well-insulated at least a few feet below ground surface. A dilemma results, though, as the more plants you have, the less direct sunlight reaches the floor. Also, little heat transfer occurs when cold air sinks to the floor. The floor is important as a temperature moderator, even with moderate temperatures. A great deal of heat must escape the floor mass before room temperature approaches freezing. This can be the difference between a plant just losing a few leaves on a very cold night and dying. In short, if you have a well-insulated wood floor, you may be worse off instead of better.

There is little doubt that storage mass directly in the rays of the sun does the best job. The key to successful storage is thermal coupling. That is, getting the heat in during the day and out again at night. Any mass will help, but the best use of storage is in a place where energy enters easily and rapidly and leaves equally and tightly coupled to the greenhouse system. For this reason, rock walls, water barrels and other storage systems which are fully illuminated by the sun tend to function best. If they do not, a significant part of the energy passes into the air. The storage system must extract it from the air before the air gets so hot that ventilation is needed. Naturally, if you ventilate, you are losing more heat that will have to be made up by a stove or other heater at night.

Clearly there are several problems in getting the storage systems into the direct sunlight. First of all, we want to grow lots of plants. If there are lots of plants (particularly nice healthy, tall, bushy tomato plants), they will shade the storage resulting in the unpleasant situation of having a greenhouse that works great as long as there are few plants in it. The second problem is one of geometry. If the sun is just _so_ high in the sky, then it will just shine _so_ far back into a greenhouse of a given height. To allow for illumination on the storage, it must be at a distance from the glazing where the sun can penetrate. Here the height of the glazing is the deciding parameter. As a general observation, the glazing height needs to be of a similar dimension as the distance from the glazing to the storage. (This is if direct illumination is expected for a reasonable amount of time.)

Also, even though the sun is in the southern part of the sky all winter, it is still significantly east in the morning and significantly west in the afternoon. So a greenhouse, which has a short east-west axis in relation to its height, may well find the sun shining on a given part of the greenhouse for a short time due to end-wall shading. From a purely practical standpoint, a successful design places the storage on the north wall; will have glazing at 45 to 60 degrees on the south wall only; will have glazing approximately as high as the greenhouse is deep; will have an east-west dimension two or more times the north-south dimension. Such designs--evolved in the sunny Southwest--allow storage to be in the direct rays of the sun most of

HUDDLING TOGETHER IN MOTHER EARTH

Presented and prepared by: Fredric A. Benedict

INTRODUCTION

Since earliest times, man has taken to the earth for shelter. Living in caves lost its popularity, but in many places in the world people still live underground. In the Loess Belt of China, ten million farmers live in atrium-like houses 25 feet under the surface. In Capadocia, Turkey, forty underground cities have been carved out of the volcanic turf. In Gaudix, Spain, 30,000 people live in apartments carved out of the cliffs. In France's Loire Valley, ancient Troglodyte dwellings have been taken over for luxury weekend abodes. Adobe brick buildings enclosed in an envelope of insulation and employing passive solar heat have been extremely successful in the American southwest. Most Navajos still prefer their snug earth-covered hogans to the white man's flimsy houses.

In addition to residential structures, commercial complexes, libraries, theatres, museums, office buildings, churches and schools have been built underground. Place Ville-Marie and Place Bonaventure in Montreal are good examples of this. They were built in and over a large open cut for railroad tracks at the central station. The complex of shops, restaurants, offices, cinemas and passageways are very popular and an excellent solution for a city center in a cold climate. The city plan for

Moscow illustrates underground commercial installations, swimming pools, warehouses, garages and other public services. Planners claim 18,000 acres of surface area would be freed for parks and sport facilities.

A lot of energy can be saved by constructing buildings underground. A Bureau of Standards study estimates that over the next 25 years the nation could save one hundred billion dollars in heating costs if all buildings were placed underground. But, people are usually put off by the idea of living this way. Even a visionary such as Frank Lloyd Wright wrote: "A house should not ordinarily have a basement. In spite of everything you do, a basement is a noisome, gaseous, damp place. From it comes damp atmospheres and unhealthy conditions...It usually becomes...a great furtive underground for the house in order for the occupants to live in it disreputably."

More recently, Charlie Wing, in his book, "From the Ground Up", takes a crack at underground space: "The cellar developed for the storage of vegetables and later was modified to accommodate bulky, coal-fired, hot air furnaces. Now people convert cellars into living space. But basements are expensive to build and are damp, lightless rooms, anyway. Thus, Wing states, there is no longer a need for basements.

The Maine Audubon Society employed a solar system in its state headquarters. A spokesman had this to say, "We had thought about digging a hole 20 feet deep and covering it over because that would really be the most efficient. Once you get down a little way, the ground is always about 50 degrees. But nobody wants to live that way."

A dwelling should not be completely cut off from sun and view. Jay Swayze of Plainview, Texas tried to sell the idea of a completely underground house at the 1964 New York World's Fair. It had a self-contained climate and environment, false windows, fake sunsets and painted panoramas. But people were not impressed. One can take advantage of the earth's heat and still provide

Solar Architecture

a view. It is easy to do on a site sloping to the south. Even with a flat site one can provide an intimate view such as from a cloister. John Barnard has demonstrated this with his "Ecology House". All rooms open to a sunken atrium. Additional advantages with this urban concept are greater privacy and reduction of street noises. An underground public school has been in existence in Abo, New Mexico for 15 years. The desirability of this space is controversial because it is completely underground. The underground school in Reston, Virginia makes more sense with its entire south wall of glass.

Advantages of Earth-Integrated Architecture
Why take the trouble to build in the earth? First, studies show that only 20% as much energy is required for heating and cooling as an above-ground structure. That is all that many people expect to accomplish with a solar system. Earth is a poor insulator but a real temperature moderator. A three foot blanket of earth around a 1600 square foot building weighs over 800 tons, with heat storage capacity of 300,000 Btu's. With such potential, heat loss from the building at night or during cloudy spells is unnoticeable. It is the "thermal flywheel effect". Soil takes on and gives off heat very slowly. (The earth reaches it peak of summer heat in November and winter cold in May.) There is a three-month lag from surface peaks.

There are many advantages to earth-integrated architecture. First of all, below ground space can be cheaper to build. This is especially true of basements where frost conditions require one to build 4 to 6-foot deep foundations anyway. Also, there is less air infiltration with an earth structure and it is more fireproof. Some buildings have been pushed below-grade for aesthetic reasons. College campuses and land-scarce downtown areas, museums, libraries, parking garages and other buildings have been built with parks on the roof. An earth building is generally less conspicuous, more organic and fits into a natural setting with less visual impact.

Andrew Davis built an earth covered house in Arlington, Illinois. Last winter he used 2½ cords of wood to heat it. As an experiment, he let his stove go out. Because of the flywheel effect, the temperature only dropped two degrees per day--going from 70 degrees to 62 degrees in four days.

There seems to be a cumulative effect when an earth building is heated. The surrounding earth gradually warms up, reaching its maximum temperature in 2 or 3 years. Davis reports that his earth is now 58 degrees. At the University of Minnesota bookstore, a large underground structure, the earth temperature is up to 65 degrees.

One of the unsung heroes of the solar movement is Wendell Thomas. His two houses were built twenty and thirty years ago but, because they involved only passive solar, they are not referred to as "Solar Houses". Yet, the back-up heating required in Thomas' houses was small compared with that of the well known solar pioneers who designed active solar houses. This was because Thomas had the foresight to minimize the surfaces exposed to the elements. He said, "The spread out ranch house is fine in a mild climate like Texas, but go two stories when building in a cold place like Minnesota." To further minimize exposing walls to the cold, he dropped the lower story down into the ground and took advantage of Mother Earth's heat. In the mountains of North Carolina, where temperature varies over 100 degrees, Thomas' house, without artificial heating or cooling, has a temperature spread of only 15 degrees from the coldest winter morning to the hottest summer afternoon (60 to 75 degrees). Thomas feels that there should not be any openings in the basement walls. This may be true in a damp region, but in a dry climate, it is desireable to incorporate south-facing windows. This can be accomplished even on a flat terrain. In a house I just built for my family, I dug a basement with south-facing windows for our children to provide a cheap and energy-efficient space.

For the same reasons that buildings do not need solar hardware

Solar Architecture 33

to qualify as solar architecture, buildings must not be underground to be called earth architecture. If a building has usable space in a basement that benefits from the earth's heat, it is earth architecture. It is all a matter of degree. A building with 3 or 4 feet of earth on its roof receives additional benefits, but may not make sense economically. The additional structure required to support 400 pounds per square foot of soil is very costly.

The sod-roofed house (elevation 8,000 feet) that I built nineteen years ago was often too cool in the summer. At 5000 feet elevation, I recommend that an earth roof have a sprinkler system to cool the house in the summer. In our climate of the high Colorado Rocky Mountains, there is a benefit from the early snows. The snow acts as insulation, so frost penetration is less significant. Therefore, the soil near ground level retains a higher temperature and it is not as important to sink the building deeper into the earth.

| Figure 1 | Elk on sod roof of house designed by Fredric Benedict |

For some, returning to the cave is difficult. For the last fifty years, the international style architects have told us that we must pierce the walls with glass: "Bring the outside in...open it up to nature and the view." It is not easy for us to completely reverse that thinking, and modern architects are slow to accept the idea that a dwelling should be first and foremost shelter from the elements.

There is a similarity between the human body and buildings as shelter. When going out into subzero weather, the prudent person will bundle up--pulling his wool cap over his forehead down to his eyebrows and leaving only part of his face exposed. Modern architects (with their 50 to 70% glass) have stripped us to the waist before sending us out in the cold... in some cases, below the waist. That is alright in Acapulco, but ludicrous in Buffalo. Architects have not really faced up to the fact that a well insulated wall is over 10 times as good as double glazing.

It was called the International style, and for 40 years architectural students were taught that the same building that works in Germany or Connecticut, will work in India and Africa. What happened when European and American architects began building their concrete and glass cubes in the Middle East? The Arabs soon discovered that their traditional way of building with mud bricks and slit windows was not so naive. Of course, a lot of glass and concrete buildings are still being built there and in other developing countries. The mechanical engineer takes care of everything--by pumping in the energy-consuming air conditioning.

I remember years ago a mechanical engineer showed me his house in Denver. He was so proud that there was not an opening sash in the place. What was ironic was that a huge broadleaved cottonwood grew just west of the house. With the shade from that tree, he could have been very comfortable with cross-ventilation and no air conditioning.

In the United States, we are still building World Trade Centers (with an energy use equal to a city of 150,000 people) and Sears Towers (equivalent to the energy consumed by the city of Rockford, Illinois). Have any of you seen an office tower going up with a windowless north facade? More likely you will see another award-winning Pennzoil Place. Its designer was recently feted by fellow architects on the 25th anniversary of his all glass house.

When the General Services Administration began to plan their experimental energy saving building in Manchester, New Hampshire, the architect designed 50% of the wall space in glass. Fred Dubin, the energy consultant, wanted that reduced to 5%. They finally compromised on 12% fenestration on the east, west and south with none on the north.

The "Edifice Complex" is widespread: Twenty years ago, I built a small house in Aspen with an underground bedroom wing with sod roof. The owners of the adjoining property were offended by such a "Non-House" and sold their lot. The house I built for my family nineteen years ago was also very inconspicuous with its sod roof fading into the woods and much of it built underground. When we decided to sell the house in 1975, it took a year to sell, perhaps because it was so inconspicuous; if someone pays expensive Aspen prices for a house, they want a house people can see! My wife too wants a house to have a face: the arched door, its nose, upstairs windows, its eyes... She says a house has either a sweet or mean expression.

Whereas some solar advocates are impressed by Pueblo Bonita (Chaco Canyon, New Mexico) with its eight hundred rooms and realize that it could be a prototype for an energy-conscious community, the majority are still pursuing the "Old American Dream:" a free-standing house on an acre of land surrounded by National Forest. Many are looking for a "different" way of building which will help them beat the system. First, it was A-frames, then Domes, Zomes, Teepees, Tetrahedrons and whatever.

GOVERNMENT, LAND-USE POLICY AND ENERGY

While building support for President Carter's energy program, James Schlesinger said that we are not going to reduce the American standard of living and we are going to go on with suburbanized homes. Can we really be serious about saving energy and continue to build 1,000,000 houses per year in the suburbs? If it is true that space heating takes 20% of our energy but transportation takes 34%, the handwriting is on the wall. We must reduce commuting by automobile. Why do several western European countries have GNP and standard of living comparable to ours but only use one-half as much energy? Mainly because they live in cities and use good public transit to get to work. Many now have cars that are used only on weekends to get into the country.

In "Life and Death of Great American Cities" Jane Jacobs blasted the sacred planning concept that concentrating people is unhealthful and degrading. She pointed out that the North End of Boston has the highest density in the city (275 dwelling units per acre) but has the lowest delinquency, disease and infant mortality rates in the city. It has an atmosphere of buoyancy, friendliness and good health. Jane Jacobs points out other attractive areas in cities that have high densities: Telegraph Hill-North Beach in San Francisco, Rittenhouse Square in Philadelphia, Greenwich Village and Brooklyn Heights in New York.

Jacobs points out that there is a vast difference between high density and overcrowding. Overcrowding is prevalent in slums of many large cities such as New York and Los Angeles. It occurs when decent housing is not available at a rent that low-income people can pay. City life under such circumstances rapidly deteriorates. Robert White of Boulder said it all:

> It is time for local governments to realize that we bear a large responsibility for inefficient energy usage and that we have a unique opportunity to change that pattern of waste.

Local governments through their building codes, and most important, their land-use regulations (or lack of them) have promoted the waste of energy and the destruction of energy-efficient central urban areas. Local governments must now, using these same tools, reverse these past policies. Over the past two or three decades, we have seen city after city permit, in fact, actively encourage tract after tract of sprawling, energy-wasting suburban housing.

Each year, we in this country build over a million single family detached houses scattered over the landscape. This is the most energy-inefficient housing in history. The staggering amounts of fuel consumed for transportation between these suburbs and employment centers, schools, shopping, etc. constitute only one aspect of their appalling energy waste.

We have allowed a very high percentage of this housing to be built to standards of construction and insulation that is laughable. Even, with acceptable insulation levels, the 1200 square feet of heat-leaking roof area covering the average single-family house still results in a monumental waste of home heating fuel.

These problems have been ignored or glossed over in the long infatuation of local governments with suburbria. Alternative, energy-efficient, higher-density, multi-family urban land uses have been actively discouraged and resisted by many local officials.

Having decided that "nice people don't live in anything but single-family houses," they then went on to make that prejudice a self-fulfilling prophecy by relegating multi-family housing (if permitted at all) to the noisiest, least attractive parts of town; saving all the choicest areas for rows and rows of monotonous, wasteful, look-alike sacred cows.

Through restrictive zoning and official resistance to the development of all housing but suburban sprawl, local government has greatly contributed to the energy crisis we face today, not to mention the decline of our central cities as attractive places to live.

We must stop promoting suburbia and start

encouraging the development of high-quality, higher-density urban areas. We can begin to make owner-occupied multi-family housing an available and respectable alternative to suburban living.

There is a tremendous opportunity, through local land-use policies, to begin to educate the public to the need for and possibility of attractive urban living.

The actions we take at the local level are going to have effects well into the 21st century. Each time we approve a remote subdivision, each time we neglect to set responsible building standards, each time we downzone a piece of property, we are directly encouraging a wasteful use of energy and resources that will probably remain for 30 to 50 years.

The energy crisis didn't happen to us, we created it and the only responsible way to deal with it is to change the policies that brought us to it.

White made these remarks at a National League of Cities meeting, emphasizing the role of local government. The federal government, with its massive super highway program and mortgage policies must share the responsibility for the sad state of affairs that exists. The freeways put vast acreages within commuting distance of the central cities and promoted the leap-frog practices of tract builders. They were encouraged to go farther out where land was cheaper to build their monotonous developments. If we now emphasize close-in, medium-density housing, the heat savings over one story single-family houses is very impressive. If it pays off in energy savings to reduce exposure to the elements by building two or three stories and sinking into the ground, think of the additional savings if a heated party wall exists on either side of a dwelling. Robert White has found that in Boulder he can build 2,000 square foot townhouses with 15,000 and 20,000 Btu furnaces. A typical freestanding one-story home would require a 100,000 Btu furnace. This remarkable accomplishment is partly due to his being both architect and builder and thus able to control the building process. Insulation is installed properly, infiltration is controlled due to thorough supervision. There are many heat-saving practices.

Solar Architecture

Combustion air for fireplace and furnace is ducted from the outside, warm air is recirculated from top floor to lower floor, and warm air registers are away from windows. Windows are concentrated on the south for direct-solar gain. The chief reason for its energy savings, however, is the use of party walls and the spreading of roof heat losses over two or three stories.

If we are willing to live in enclaves, then we will be able to do something about the extremely wasteful, auto-oriented society which we are still building. It is only through concentrated densities that we can manage without the automobile and best utilize mass transportation. The enclave would not have to be immense, as the 700,000 inhabitants proposed by Soleri in the Arizona deserts. Something on the scale of Chaco Canyon would be more humane. We could huddle together surrounded by nature to conserve energy and best utilize our natural resources.

BIOGRAPHICAL SKETCH

Fredric A. Benedict is an architect in Aspen, Colorado specializing in the architectural design and land planning of recreational areas in the Colorado Rocky Mountain area. Mr. Benedict is also a member of the Board of Advisors of the Roaring Fork Resource Center.

KEEPING A COOL HEAD AND WARM FEET

Presented by: Keith Haggard
Prepared by: Keith Haggard, Barbara Francis, Larry Palmiter

Human comfort is a topic that dissolves as one approaches it. The best one can do is to take it by the hand, like the leprechaun, and wait while it changes shape before asking where the gold is kept. Is the man in the beat-up sweater, chopping wood on a morning when he can see his breath, more comfortable or less comfortable than the air traffic controller in shirtsleeves at an even 70 F? Comfort is a matter of human perception and social emphasis, likely to yield to discursive circling rather than direct attack. I doubt that we will be spared the spectacle of direct attack--comfort specialists and comfort analysts, Ph.D's in Comfort, Comfort Colloquia and, inevitably, Federal Comfort Standards--but for the moment, we will do better to put our feet up on the conference table and relax, exchanging notes to see if we're looking at the same mirage.

We have to begin with the assumption that we are talking about the comfort of people in buildings and eliminate the unsettling possibility that people are more comfortable out-of-doors. That possibility is worth retaining and contemplating in our individual leisure, while we're mowing the lawn perhaps, but we should take it for granted for now that our society needs buildings, that we need our society and that we should work, as a consequence, to make our buildings more comfortable. Whether or not

we are actually comfortable in a given building depends on a
variety of factors ranging from its shape, size, texture, succession in plan, smell, temperature and lighting level, to our
own health, history, purpose in being there and opinion of its
owner. Nevertheless, we will have to narrow the topic further
and take comfort in buildings generally to mean thermal comfort,
partly because we know more about thermal comfort than we do
about the rest and partly because our shared immediate interest
is design with climate, which is primarily thermal design. Comfort and climate invariably interlock; we are designing <u>with</u>
climate <u>for</u> someone, and his perception of our success is largely
its measure. Comfort and climate are two of the perpetual fresh
water springs of inventiveness in design. They are always new.
The last word will never be said about them. Buildings certainly can be designed as if they had no occupants and no real setting in the climate, but they ought not to be and life will be
more interesting if they are not.

One of the first, and few, firm points to be made about thermal
comfort is that it's relative. Standards of comfort vary from
time to time and from place to place--probably everyone knows
that the English think our rooms are overheated. If you look
at some of our own older mechanical engineering books, you will
see recommended indoor design temperatures consistently in the
60's, with higher values only for "sick rooms" and other unusual
circumstances. The indoor design temperatures have been steadily
creeping up over the years, at least in the winter; the summer
ones remain low and are now working their way through the 70's.
In the most recent edition of the ASHRAE Handbook of Fundamentals, the recommended winter indoor design temperature is 75 F.
This figure is justified with the remark that people are dressing
differently these days and currently go around clad in only 1/2
clo in the winter; however, in the summer, when the indoor
design temperature is 65 F, most of us carry a jacket in self-defense.

Solar Architecture

The clo, if it's unfamiliar to you, is the engineer's measure of clothing. The clo scale has the birthday suit for its zero point, with other values ranging from 1 (light) to 4 (heavy). The upper limit is reached when your clothing prevents you from moving at all, like a stuffed animal. We are anxious to assign numbers to complex sets of conditions like clothing, presumably in order to deal with them more realistically, but the numbers, once assigned, tend to take on a life of their own. We forget that we put them there for our convenience. We treat them as real in themselves, rather than as handmade nets casting about for reality in a choppy sea. When the numbers let us down and lose touch with reality, so do we.

This is not meant to denigrate numbers. In many cases, numbers are very convenient, correctly assigned, and eminently useful. We can analyze a climate we have never lived in through numbers and need not wait for successive local generations to develop suitable building forms by trial and error. We can trade possibilities through time and across continents without leaving the library. The process is risky and does not always work out well, but it's possible as long as the numbers have meaning. Inches of precipitation, daily temperatures (highs and lows) and watts per square meter of sunshine have meaning, being truly measurable and by and large beyond politics.

The real problems arise when we edge from the measurable to the perceived. When we come to human perception or behavior and its measurement, even just to one of its fringe issues (like how people in buildings are "usually" clothed), we have to take some pains to preserve significance. Our methodology, how we came to state what we say we know, becomes crucial if our nets of numbers are to catch anything at all. Worse yet, at some point in describing perceptions with numbers, we have the right, or perhaps even the responsibility, to ask if the numbers really ought to be what they are. That's a new ballgame. No one seriously asks if the R-value of a material or the efficiency of a collector really ought to be what it is measured to be--not in the same sense. The fact that measured R-values and efficiencies vary according to who measures them is a cause for concern perhaps, or a sceptic's delight. It is always possible to say that R-values ought to be higher or collectors more efficient from a design standpoint, but it isn't the same as asking whether or not people "ought" to go around in 1/2 clo in the winter, if that's what they want to do. That is a cultural "ought", and a tricky one. If standards of comfort are relative, by what standard do we evaluate them?

The problem is knotty enough if we simply observe that standards of comfort and perceptions of comfort are different. It becomes knottier still if we consider that perceptions of comfort are different because standards of comfort are different. For better or for worse, it seems that we learn to be comfortable. These learned perceptions are similar to an autobiographical paragraph from a forgotten source: the writer confessed to an entrenched sneaking conviction that waking an Englishman up in the middle of the night would catch him with his accent down. Taken sufficiently by surprise, he would drop his silly pretensions and talk straight, like a fellow from Maine...well, maybe Michigan. That is not true, and even the writer knew it, but it showed how he felt about learned accents different from his own. They were suspect. Anyone who grew up "the real way" would have the good sense to talk like he did. Many of our perceptions, including our perceptions of comfort, appear to be learned, like our accents, out of a very broad range of human possibilities. The fact that they are learned, however, does not make them less real. If a man claims to be uncomfortable in circumstances that make a Sherpa mountaineer feel quite at home, he is not lying, nor is the mountaineer lying if he finds the other's normal conditions intolerable. We have learned to feel at home in different places. Relatively different standards have ended in genuinely different perceptions.

It is tempting to think that we might learn to perceive something different. The Sherpa mountaineer and the other man might learn to change places and feel comfortable in each other's surroundings. It might be hard, since some kinds of learning seem to get harder as we get older and more obstinate, but it might be possible...and it might be a good idea for at least one of them. That is the question of "ought." "Ought" we to be comfortable in the particular way that we perceive comfort? This is a question we can ask of any civilization, wherever history beaches us, and not just of our own, or more accurately, not just of yours and mine. It is not a moral question, or at least does not have appropriate moral answers. This is not to suggest that some standards of comfort--like yours, for example--are decadent and immature, while others--like mine, for instance--show a high level of psychic accomplishment. The problem is one we face almost anywhere we turn these days, and it is uniquely ours only by its scale. The problem is: how are we to evaluate the behavior, including perceptions and their expression, that we learn and teach? By what rule do we distinguish "good" responses from "bad" ones? The only difference unique to our own wrestlings with this particular angel is that they do not take place within the confines of an obscure cult, whose conclusions will be merely curiosities beyond its edges. Our problem, which dawns on us with increasing certainty and dismay, has been posed to a world society capable of deciding it once and for all for the species. It is actually as if we had to choose between good and evil <u>right</u> <u>this</u> <u>minute</u>, for centuries to come. We seem no better equipped to do that than any other group has ever been.

In our context, the question presents itself as follows: Is it "good" to be socially conditioned to feel "right" or "normal" in certain sets of environmental conditions rather than others? Is it good for the organism? Is it good for the conditioning society? Is it good for the species? We have few answers; we have few clear ideas where to look for answers; and, we have few ideas about what to do with answers we might find. It is tempting to frame the question in terms of health and longevity. The healthiest life you can lead, the life that consistently gets you past the hundred mark, should you live past five to begin with, is the life of a villager in the Andes or the Urals, somewhere up above 10,000 feet. If you will take up that form of life, you can apparently even drink and smoke and live to tell about it at one hundred and one. Why might that be? Is it the composition of the air or its pressure? Is it the diet? The exercise? The community size and pace? Is it genetic? Health itself is a rather arbitrary standard--surprisingly large numbers of people have literally elected not to live very long for one reason or another, while others have hung on indefinitely in ways we might find appallingly pointless. But decisions cannot even be made in terms of health until we know what the variables are. We cannot say that our office buildings should be at 70 F, rather than 80 F or 50 F, or even whether or not we ought to be _in_ office buildings, until the returns are in.

"Life consists of acting on insufficient evidence," says William James, but some forms of evidence are less sufficient than others. This has been a roundabout route to the point, which is that you cannot simply walk into an office building, ask people if they are comfortable, plot their responses on a chart and make that a professional standard, presumably as useful as a temperature record and potentially worth enforcing by code. Even if you could get honest answers, which is largely a matter of how the questions are put, the answers would lack meaning. Why are people comfortable, or why do they say they are, and is their comfort a "good" thing, either for them or for us?

The inadequacy of the survey approach is recognized in two of the most technically promising, and certainly most purposeful, branches of research into human performance in varying thermal conditions, which we can distinguish as "extreme performance" research and "optimization" research. It would be wrong to say that their topic, like ours, is comfort, since comfort is usually out of the question, but they do have something to say to us, if only off-hand.

Extreme performance research tries to identify the outer limits of human ability in very unpromising, potentially injurious,

Solar Architecture 49

makes people happy, while a temperature swing of about 8 F or so every 16 minutes makes them faster at adding up numbers. All of these findings are limited, but they may at least show us a direction to take. Wyon himself is drawn to infer that we need better feedback loops in our mechanical systems, allowing considerably more occupant control of variables like temperature and air change. He contrasts this higher level of user control with what he calls the "zookeeper" approach of current practice, in which environmental conditions are specified by the designer and are presumably maintained as specified by the mechanical equipment. Whether or not our buildings actually work as they are designed to work is debatable, and well worth debating. As a slight exception to Wyon, it seems that the apparent human worth of things like temperature swing and air change simply implies that we can legitimately let the <u>climate</u> control a bit more than we do now.

How buildings actually work thermally is the present preoccupation of an increasingly important figure in building research, Dr. J.L. McGrew of Denver. McGrew is a measurer and has come up with some numbers in real contexts that are literally startling, if you are capable of being startled by heat transfer. McGrew is a veteran of the aerospace program, where he did human heat transfer analysis for the Apollo project, and his vantage point is essentially physiological. He feels, reassuringly, that we can save energy by making ourselves more comfortable in our buildings by fitting them better to human thermal perceptions. Before confronting McGrew, however, we ought to consolidate our ground with a short review of the basics of human heat transfer.

Figure 1 sums it up. We work from the inside out to maintain a body temperature characteristic of our species, despite considerable differences in age, weight, sex, activity level and so on. It is really quite a remarkable feat when you think about it and it takes a lot of complicated chemical equipment to do it. In general, we transfer heat like any other real objects in space: we trade radiation with the surfaces around our own

HEAT GAIN

HEAT GAIN:
- INTERNAL METABOLIC GAIN.
- ACTIVITY - RELATED GAIN.
- CONDUCTION/CONVECTION FROM HOT AIR & SURFACES.
- RADIANT GAIN FROM SUN AND OTHER HOT SURFACES.

RESPONSE:
- DECREASED ACTIVITY
- INCREASED BLOOD FLOW
- EVAPORATIVE REGULATION

EFFECTS:
- RAISED SKIN TEMP.
- DECREASED EFFICIENCY
- RAISED BODY TEMP.

HUMAN HEAT OUTPUT

- SLEEPING — 250
- SITTING QUIETLY — 400
- SITTING eg. TYPING — 450-550
- SITTING eg DRIVING 500-600
- STANDING eg. BENCH WORK 550-650
- SITTING HIGH MOTION — 600-800
- STANDING - SOME WALKING — 650-750
- STANDING - MODERATE WORK 750-1000
- WALKING - SOME LIFTING — 1000-1400
- INTERMITTENT HEAVY LIFTING — 1500-2000
- RUNNING — 1500-2200
- HARDEST SUSTAINED WORK 2000-2400

HEAT LOSS

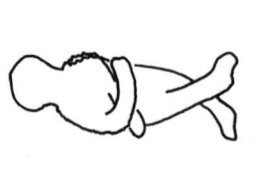

HEAT LOSS:
- CONDUCTION/CONVECTION TO AIR BELOW SKIN TEMP.
- RADIANT LOSS TO COLD SURFACES.
- CONDUCTIVE LOSS TO COLD SURFACES CONTACTED
- EVAPORATIVE LOSS THRU LUNGS.

RESPONSE:
- INCREASED ACTIVITY
- DECREASED BLOOD FLOW
- MUSCULAR TENSION
- SHIVERING

EFFECTS:
- REDUCED SKIN TEMP.
- REDUCED BODY TEMP.

Figure 1 — Human Heat Transfer

Solar Architecture   51

exposed surfaces; we conduct heat back and forth with whatever touches us; and, we exhaust moisture for evaporation from our skin and lungs. Human heat transfer is usually a heat loss from our point of view and that is actually its purpose, if it can be said to have one. Human heat transfer disposes of the extra energy we generate by eating, living and not being perfect engines. Under ordinary circumstances, human heat transfer is about 2/5 radiative, 2/5 convective (as we warm the air around us and it moves away) and 1/5 evaporative. When conditions heat up, evaporation gets more important. It gets crucial, in fact. That is true because, as one authority puts it, "In cold conditions, the skin temperature is higher than the air temperature, while in hot conditions, the gradient is reversed." Increased evaporation, or sweating, is our very creative response to that stunningly obvious state of affairs. It would ordinarily be impossible to lose heat to an environment hotter than we are, and we solve the problem by changing its terms, by producing more moisture for evaporation. Every ounce of moisture evaporated clears its immediate vicinity of about 60 Btu's of heat. "Dry" heat is often thought to be much more tolerable than humid heat because we sense our own evaporative cooling, which takes place more readily in drier air.

How we sense things like that is a fascinating physiological puzzle that could well occupy a professional lifetime. The process seems to start with the skin on our hands and feet and the backs of our necks. These are nerve centers and they give us readings for use in many mysterious processes. In the context of thermal performance, they seem to inform us as to the ambient temperature, the relative humidity, the wind speed and the radiant conditions around us. The information comes all in a piece and is usually designated as "effective" by the engineer. The effective temperature of a room is the one that it seems to be at, rather than the one cranking along on the strip chart recorder. The idea that what we might call "effective comfort" can be achieved at rather low temperatures by various ploys is the hopeful one that McGrew has to offer.

When the environment, as sensed by the skin, fails to cooperate and make us effectively comfortable, the body acts. Some of the activity is rather gross: when it starts to get a little too hot or too cold, people begin to do visible things, often unaware. They hug themselves and jump a bit, they huff on their hands, they loosen their ties, they spread themselves out... without really seeming to notice what they are doing. The classic demonstration accompanies the traditional school of architecture lecture on human heat transfer: the lecturer has someone gradually advance the room thermostat as he talks, until someone finally remarks how hot it has become. That is usually long after the students are half-undressed, sprawled in their seats and fanning themselves (some are also asleep by then, but that is the control group). We are very adaptable beings over the short term, especially when distracted by a task like taking notes or listening. That we may be too adaptable for our own long-range good is the main theme of another interesting thinker, Rene du Bois.

When activities like moving around do not suffice, we take more sophisticated steps, again through the sensitive skin of our hands, feet and napes. These interface areas can change as thermal connections to our surroundings. More or less blood can flow just beneath them, causing them to transfer more or less heat to the environment as a result. One might think that we would try to provide more heat to our hands and feet when they claim to be cold, but the reality is the opposite. When our extremities claim they are cold, we start to shut them off. Their blood vessels contract and less expensively heated blood flows through them. This is the basis of frostbite. The frostbitten areas have simply been written off by the body on the understanding that it can live without toes, but it cannot live without a certain internal temperature. The body has a set of built-in values all its own which put survival ahead of comfort. We are uncomfortable as we edge our way towards frostbite, but it is simply not worth the cost to keep our toes warm until our heart gives up.

When we are too hot, on the other hand, more blood is delivered to the skin. The vessels dilate, the skin conductivity goes up, and more heat is delivered faster to the environment. With a normal skin temperature in the 85 to 95 F range, we can successfully lose heat to any environment cooler than that. When surroundings are hotter, or when we are exposed to direct radiant gain from something like the sun, or when we are generating an extraordinary amount of heat in the course of heavy work, we start to sweat and lose heat by evaporation. Direct solar gain is the other side of the coin in the dry heat/humid heat debate. Hot arid climates are clear ones and the desert sun can be enervating. Only mad dogs and Englishmen went out in it according to the old song, in odd corners of the Empire where behavior acquired in London was emphatically, even arrogantly, maladapted.

Once the designer has dragged himself, one way or another, through the basics of human heat transfer, figures like our second one begin to become interesting. Figure 2 shows the conventional psychrometric chart with the things an air conditioning engineer would need to know left out. We respond to relative humidity to the extent to which the air is saturated, rather than to the absolute amount of water in the air. We seem to be happiest at about 45% relative humidity, although that figure is difficult to make use of. If the air is very humid, we object because our skin feels clammy. If the air is very dry, we object because our mouths and sinuses feel dry. Between the two extremes, however, we seem rather insensitive.

I doubt that most of us could say with any certainty whether the ambient relative humidity is 40% or 50% of 60% or perhaps even 70%. The psychrometric chart is primarily helpful to the designer of walls and other building surfaces in cold climates: it illustrates where in the wall the dew point occurs. It is also useful when analyzing an unfamiliar climate to determine whether or not evaporative cooling will work. A difference of 15 or 20 F between dry bulb and wet bulb temperatures suggests that it will be useful to get some water into the air. Most of the non-

Figure 2: Psychrometric Chart

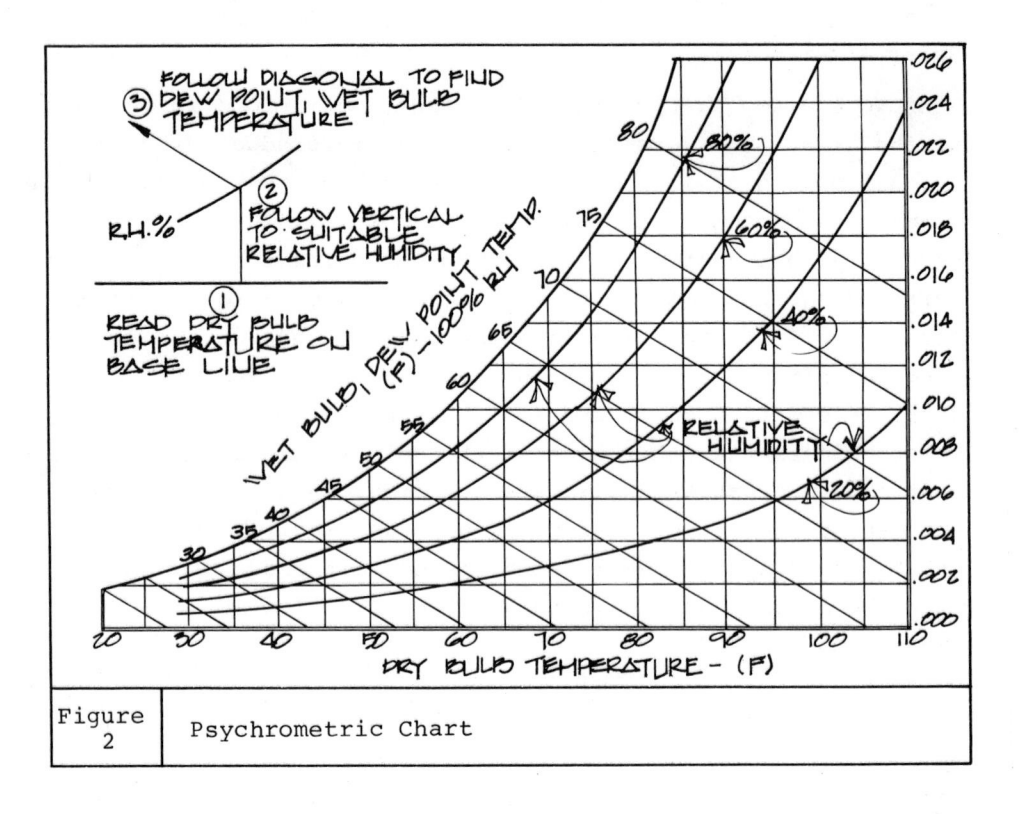

mechanical methods you can think of for doing that are probably already standard practice somewhere in the world--or were, once. Anything you can do to facilitate evaporation promotes the cooling process: increase the surface area of the water available by using shallow ponds, increase turbulence by using moving streams, spray the water through a fountain, allow it to transpire through dense well-watered foliage, or let it cover the surface of slightly porous storage jars. Turning an air flow over the water source will also help and is, or was, also common practice somewhere.

Air flow inevitably brings to mind Victor Olgyay and his collaborator Alandar Olgyay, because the Olgyay classic work, Design With Climate, includes such fine graphic studies of air flow around and through buildings. According to Olgyay, air flow becomes perceptible at about 50 cubic feet per minute and becomes uncomfortable at about 200 cubic feet per minute. Can a human really sense what that means? Probably not, but probably a human can detect a wind chill, whose effects are shown in Figure 3,

Solar Architecture   55

taken from the World Almanac, along with the poetic NOAA designations of the winds and breezes the wind speeds correspond to. It is easier to gauge a 5 mile an hour wind if you think of it as a "light breeze."

Figure 5 was contributed by Dr. Eldon Boes of Sandia Laboratories in Albuquerque. It shows how we respond to our radiant environment. "Mean radiant temperature" (mrt) refers roughly to the average surface temperature of all the surfaces around you in a room: interior walls, windows, ceiling, floor and furniture. Each pair of numbers on the table, top line and bottom line, describes the same effective sensation, which is that of

Wind Speed in Mph (1 Mph = 88 ft³/min)	Perceived Dry Bulb Temperature, Fahrenheit												
	0	+35	+30	+25	+20	+15	+10	+5	0	-5	-10	-15	-20
	5	+33	+27	+21	+16	+12	+7	+1	-6	-11	-15	-20	-26
	10	+21	+16	+9	+2	-2	-9	-15	-22	-27	-31	-38	-45
	15	+16	+11	+1	-6	-11	-18	-25	-33	-40	-45	-51	-60
	20	+12	+3	-4	-9	-17	-24	-32	-40	-46	-52	-60	-68
	25	+7	0	-7	-15	-22	-29	-37	-45	-52	-58	-67	-75
	30	+5	-2	-11	-18	-26	-33	-41	-49	-56	-63	-70	-78
	35	+3	-4	-13	-20	-27	-35	-43	-52	-60	-67	-72	-83
	40	+1	-4	-15	-22	-29	-36	-45	-54	-69	-69	-76	-87

Figure 3   Wind Chill

Designation	Mph	Designation	Mph
Calm	0-1	Near gale	32-38
Light air	1-3	Gale	39-46
Light breeze	4-7	Strong gale	47-54
Gentle breeze	8-12	Storm	55-63
Moderate breeze	13-18	Violent storm	64-73
Fresh breeze	19-24	Hurricane	74-up
Strong breeze	25-31		

Figure 4   Official Designations of Winds

being in a room whose air temperature and mean radiant temperature are both at an even 70 F. Each pair means the same thing, effectively. The last pair on the right, for example, 91 F air temperature and 55 F mrt, goes a long way to account for the agreeable performance of adobe houses in warm, dry climates. The temperature swings considerably from day to night in such climates and the heavy houses track the swing at a considerable distance, requiring several hours to register a change. As a result, surfaces like the shaded inner sides of thick mud walls tend to be coolish by day and warmish by night. The house is pleasant, even as the air temperature peaks in the afternoon.

The pairs towards the left end of the chart, however, are the fascinating ones. One of them says, for example, that if you can maintain a mean radiant temperature of 85 F across the surfaces around you, the air temperature can drop to 49 F and you will not notice it. It will not be a question of keeping a stiff upper lip in order to save energy, which goes primarily into keeping the air temperature up. Instead, you will not even notice the air temperature, or you will notice it, but think that it is 70 F. In this one case, what you don't know about the environment almost certainly won't hurt you. In fact, if du Bois and Wyon and the rest of our thermal naturalists are right, it will actually be good for you. You will spontaneously express comfort--hurray!--as the air temperature changes a bit, and you will be able to add up numbers like a whiz.

Unfortunately, it would not be a simple matter to keep the mrt of a house at 85 F in the winter while the air temperature held still at 49 F. Nevertheless, the idea gives us something to work with. If we can raise the radiant temperature of the environment, we can reduce the air temperature and never know it,

Air Temperature	49	56	63	70	77	84	91
Mean Radiant Temp.	85	80	75	70	65	60	55
Figure 5	Equivalent Air and Radiant Temperatures							

Solar Architecture

except when we pay the reduced utility bills. The intricate round-robin of comfort, radiant temperature, air temperature, energy use and costs brings us back around to J.L. McGrew, our measurer and physiologist.

McGrew's central hypothesis, supported thus far by his energy audits of real buildings, is that the major energy activity of the small American building, including the house, is heating the outdoor air as it passes through. The energy we use with such malicious abandon, as our critics would have it, does not work mainly to make us profoundly comfortable or to provide us with hot water, cooked food and cold drinks, as our utility companies would alternately have us believe. Instead, the energy, in the form of heated air, goes up and out the flues of colosally inefficient combustion equipment and up and out the damperless vents in kitchen, laundry and bath. In the invisible wake of the traveling air, our floors are cold, our rooms are drafty, our feet are uncomfortable and our bills are staggering. We can turn up the thermostat to no avail: in too many cases, we are only heating the air at the ceiling by doing that. Worse yet, we can double the R-value of the building skin, also to no avail: much of the energy is not going through the skin, but through the holes.

McGrew's central prescription, on the basis of what he's found so far, is first to clean up our equipment, both literally, in the basement, and theoretically, on the drawing board; second, to close up the holes, making infiltration controllable and useful at the living level; and, third, to cool the ceiling and heat the floor by moving the air around. McGrew uses ducts with small fans in the bottom to accomplish the third purpose. By moving about 1/20th of the room air every minute--a trivial sum in cubic feet--he can pull the ceiling temperature and floor temperature to within a degree or two of each other, a far cry from the stratified 15 F he encounters too often in real rooms, and reduce energy required for heating by about one-half. The resulting warm floor has a built-in bonus. It makes our feet comfortable, and when our feet are comfortable, we are comfortable, thanks to the way we sense temperatures.

We will be hearing a great deal more about comfort and energy in the months to come. As we become more serious, we will hopefully be spared the insipid banalities of the "Paper House," guaranteed to perform well on paper, and the "Perfect Product," guaranteed to save you 50% of your energy, no matter how you now use it. As our attention to thermodynamic detail, unwarranted in an age of cheap energy, improves, so will our understanding. As our cleverness increases, so will our comfort. If we decide in the end to go back outdoors, we will do so with something to think about.

BIOGRAPHICAL SKETCH
Keith Haggard is the Executive Director of the New Mexico Solar Energy Association and has been very active in solar activities for several years. Mr. Haggard is also a Board Member of the Roaring Fork Resource Center in Aspen, Colorado.

PASSIVE SOLAR HEATING OF BUILDINGS

Presented by: J. D. Balcomb
Prepared by: J. D. Balcomb, J.C. Hedstrom, R.D. McFarland

INTRODUCTION

Solar gains through windows, walls, modified walls, skylights, clerestory windows and roof sections provide an opportunity to dramatically reduce the total heating energy requirements of a building. When the thermal energy flow is wholly by natural means, such as radiation, conduction and natural convection, and when solar energy contributes more than half of the total outside energy requirements, then the building is referred to as a passive solar-heated structure.

Passive solar heating works very well. This has been demonstrated time and again in a wide variety of buildings located in a wide variety of climates. The occupants of these buildings testify to their comfort, to the ease of their natural operation, and especially to their low fuel bills. A principal problem, however, has been the lack of a quantitative basis for incorporation of the basic concepts into architectural design.

The ERDA Los Alamos Scientific Laboratory has been evaluating passive solar heating for one year under the cognizance of the Division of Solar Energy Heating and Cooling Research and

Development Branch. The purpose of the two-year LASL program is to provide the needed quantitative basis for design.

Test rooms have been set up at Los Alamos to study the behavior of passive, solar heating elements under carefully controlled conditions. One year of test data have been obtained on a pair of test rooms which utilize thermal storage walls concepts, one with cylindrical water storage tubes and the other with a thick masonry wall. The storage walls are located behind a vertical, double-glazed wall. During the mid-winter months each of these rooms has an average inside temperature 60 to 70 F above the ambient termperature. The temperature histories in these rooms are very accurately predicted by simulation analysis techniques developed at LASL.

Eleven different buildings have been instrumented in order to study passive solar heating elements within these buildings. These are the Doug Kelbaugh residence in Princeton, NJ; the Benedictine Monastery Dove Publications Building in Pecos, NM; four small buildings at the Ghost Ranch in New Mexico; the Bernardo Chavez solar greenhouse in Anton Chico, NM; the Santa Fe First Village Unit #1 and the residences of Bruce Hunn, Carl Newton, and Tom Shankland in Los Alamos, NM. Data from these installations will be used to validate the simulation analysis technique.

A comprehensive simulation analysis computer code has been written to predict the performance of passive solar heated buildings. The code has been partially validated against testroom results and has been used to predict the performance of numerous building geometries. The results of geographic studies done with this code demonstrate that passive solar heating systems can be expected to work effectively in all U.S. climates.

TYPES OF PASSIVE SYSTEMS
The first and simplest type of passive system is the "direct gain" approach in which one simply has an expanse of glass, usually double glass, facing south. The building should

Solar Architecture 61

have a considerable thermal mass, either a poured concrete
floor or a massive masonry construction with insulation on the
outside. The building becomes a live-in solar collector. The
characteristic sun angles result in a good situation since the
south face is exposed to a maximum amount of solar energy in
the cold winter months when the sun angles are low, and a minimum amount of solar energy in the summer when the sun angles
are high. Thus, the basic seasonal characteristic of control is
automatic. An example of this direct-gain system is the Wallasey school.[1] This is certainly the largest passive solar
structure in the world and one of the first. It was built in
1962, but it is little known. It is a large building of concrete construction with 7 to 10 inches of concrete forming the
roof, the back wall, the floor and side walls, with 5 inches of
expanded polystyrene as the insulation outside of that. The
solar wall is an expanse of glass, 27 feet tall and 230 feet
long, facing south. There are two sheets of glass. The one on
the outside is clear and about 2 feet inside of that is a diffusing layer of glass. It is a figured glass, so called, which
refracts the sun's rays, so that it irradiates the roof and the
floor fairly uniformly. This structure is heated to about 50%
by the sun; the remaining energy for heating the building comes
from the lighting and from the students. The auxiliary system
which was originally installed has not been needed. The school

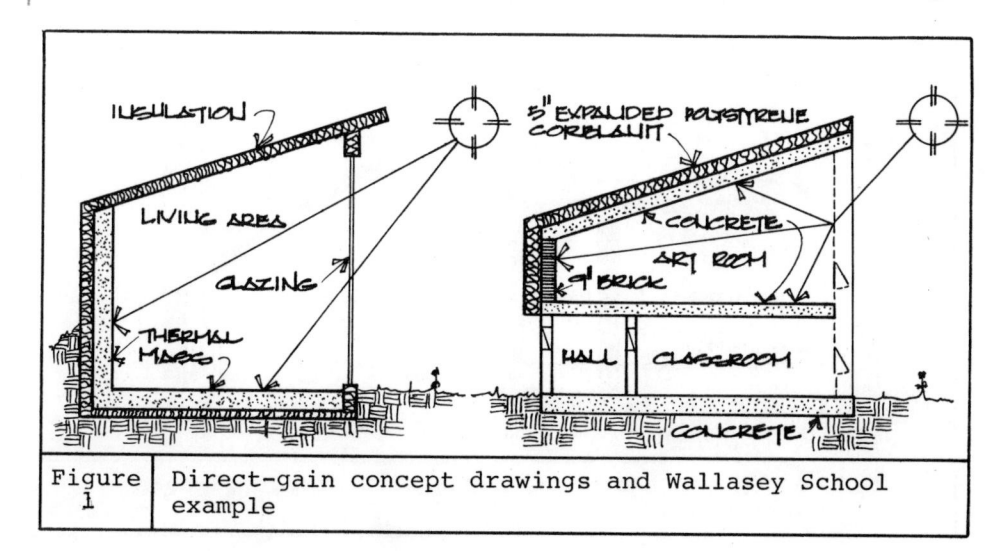

Figure 1	Direct-gain concept drawings and Wallasey School example

is located in Liverpool, England, near the sea, at a latitude of 53° north.

Another direct-gain system is the David Wright house in Santa Fe, New Mexico which uses adobe or earth brick construction with insulation on the outside and has a system of shutters which drop down at night when needed to reduce heat losses.

The direct-gain approach can take more complicated forms as shown in this sketch by Mark Chalom with the use of clerestory

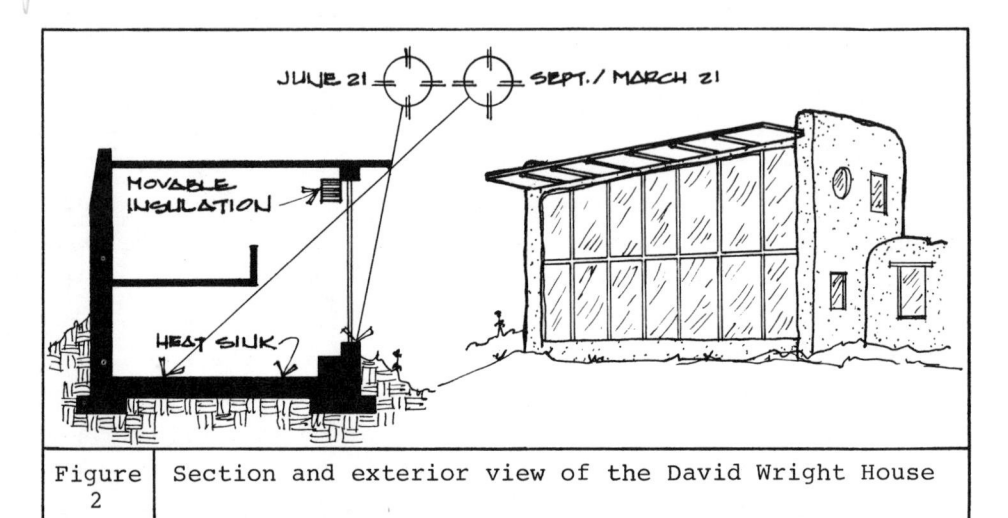

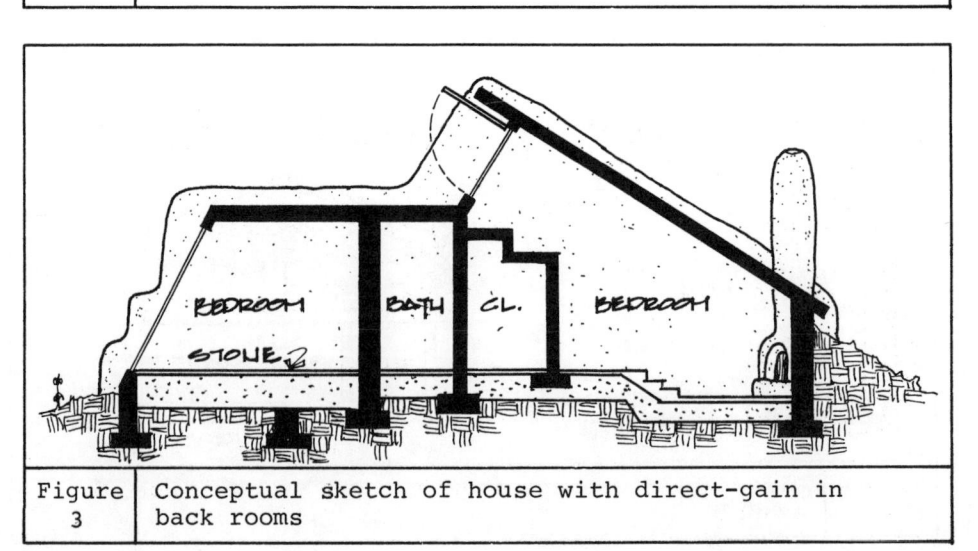

| Figure 2 | Section and exterior view of the David Wright House |

| Figure 3 | Conceptual sketch of house with direct-gain in back rooms |

Solar Architecture

windows or sky lights to provide energy in back rooms.[2]

The second type of system is the "thermal-storage wall" in which the thermal storage is in a wall which blocks the sun after it comes through the glazing, and stores the heat energy. The wall in this case is usually painted black or a dark color to be a good absorber. It can be water in containers or masonry. A small section of thermal storage wall was used in the Wallassey school discussed earlier.

An example is the Steven Baer house in Albuquerque, NM in which the thermal-storage wall consists of 55 gallon drums filled with water and laid on their sides for massive-thermal storage. A system of movable insulation is used in which a door forming the south wall lowers during a winter day to allow the sun in and actually reflect some sun onto the storage. There is a single layer of glass. The door can then be

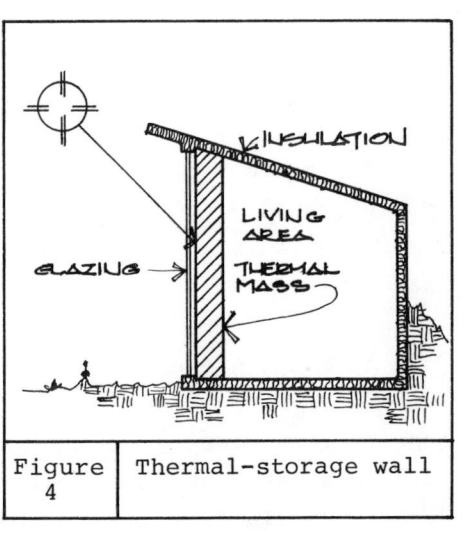

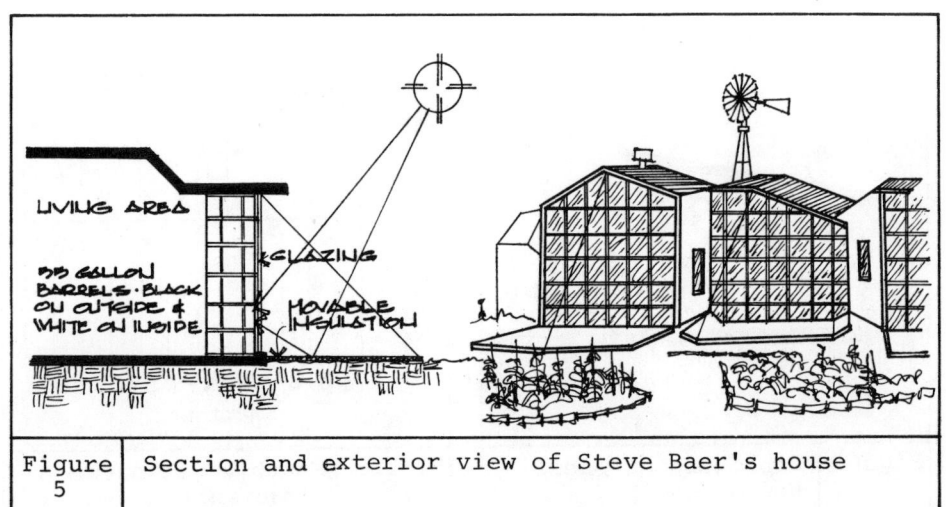

Figure 4 | Thermal-storage wall

Figure 5 | Section and exterior view of Steve Baer's house

raised at night to reduce heat loss.

A very well-known implementation of the storage wall concept is the Trombe house in Odeillo, France in which the wall is concrete.[3] In the houses that were built in 1967 the wall is about 2 ft thick. The primary mechanism for heating the house is by radiation and convection from the face of the wall with the thermal energy diffusing through this thick wall. About 30% of the energy is by a thermocirculation path which operates during the day only by natural convection with ports at the bottom and top. Some data taken on this system for a period of four days in December of 1974 are for situations in which the ambient temperature is only slightly above freezing, and there are two sunny days, a cloudy day and then another sunny day.[4] The outside surface of the concrete heats up to about 140-150 F during the day. The inside temperature remains at about 85 F, fairly uniform, providing radiant and convective heat to the room.

The data taken over a period of one year indicate that about 36% of the total energy incident on the wall is effective in heating the building during the winter months which is typical of a good active solar heating system. About 70% of the total thermal energy required by the building (which is controlled at a

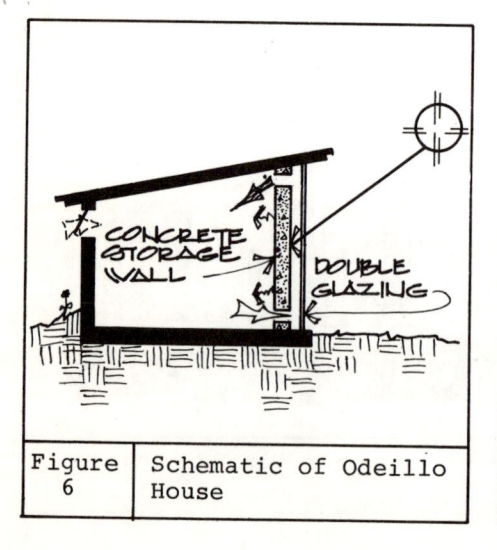

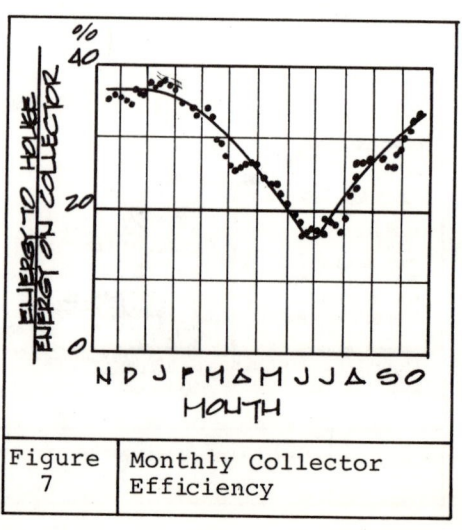

| Figure 6 | Schematic of Odeillo House |

| Figure 7 | Monthly Collector Efficiency |

Solar Architecture 65

temperature of 68 F was provided by solar energy over this one-year period.

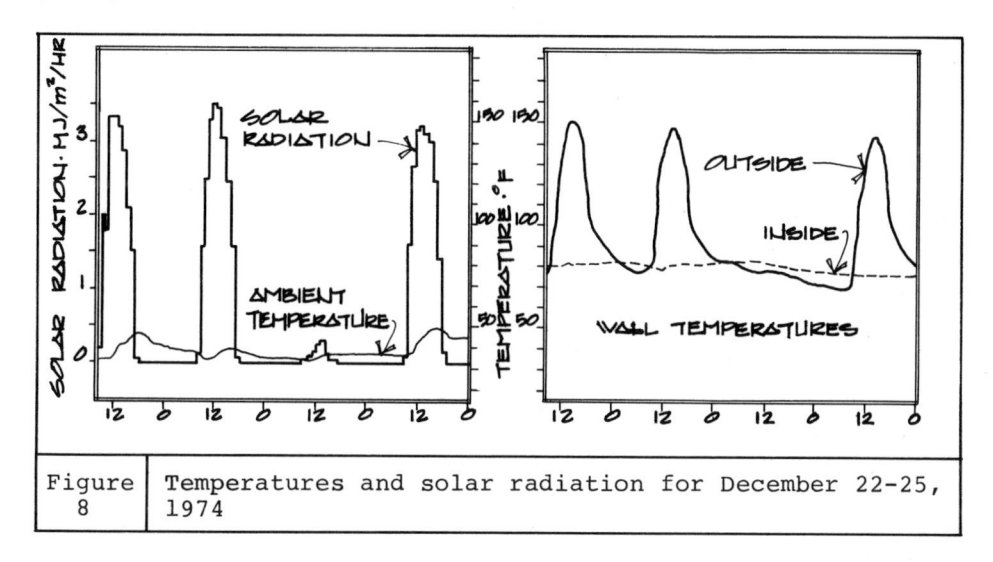

| Figure 8 | Temperatures and solar radiation for December 22-25, 1974 |

A mix of these two concepts is the solar greenhouse in which one builds a greenhouse onto the south side of a building with some kind of thermal storage wall between the greenhouse and the house. The temperature in the greenhouse does not require very good control as long as the plants do not freeze. Solar energy provides, typically, all of the heat required for the greenhouse, as well as providing substantial energy for heating the house.[5]

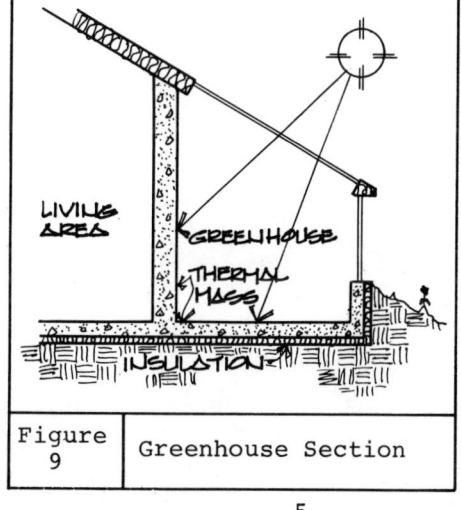

| Figure 9 | Greenhouse Section |

The fourth type of design is the roof pond in which the thermal storage is in the ceiling of the building. In this case, movable insulation is needed because the sun angles are altogether wrong. The sun provides large inputs in the summer and small inputs in the winter. It is a good natural cooling system because by using movable insulation one can take advantage of night-time radiation. This concept has been implemented in the Atascadero, California house which uses the Harold Hay skytherm concept. A system of insulating panels on the roof is used which slides back and forth on tracks.[6] The water bags are left exposed to the sky during the day in the winter and during the night in the summer. This provides heat input in the winter and heat loss on a summer night. The insulation is put in place during a winter night to conserve heat and during a summer day in order to exclude the sun which is reflected from the top of the white panels. This system has worked very well and has operated without any backup in the rather mild climate of Atascadero, providing good thermal comfort in a small building.

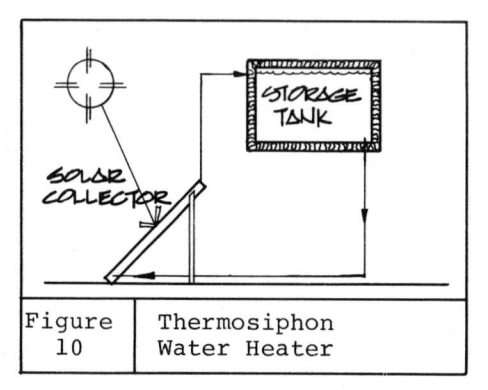

The fifth type of passive system is a natural convective loop. The classic thermosiphon water heater fits into this category. Natural convective loops which use air as the heat transport have been built and work well. A good example is the Paul Davis house in Corrales, New Mexico.

Figure 10 Thermosiphon Water Heater

MATHEMATICAL SIMULATION ANALYSIS

Thermal network analysis techniques can be used to predict the performance of passive solar-heated buildings. The building temperature state is characterized by six to eleven readings at various locations--such as air temperature, surface temperatures, glass temperatures and the temperature of various thermal storage materials at various depths. Energy balance equations are set down for each location accounting for thermal energy transport by radiation, conduction and convection, energy sources from the sun, lights, people and auxiliary heaters and sensible thermal-energy storage in the material. The temperature history of each location is then simulated by solving these equations for given inputs of solar radiation, ambient air temperature and wind. A one-hour time step is used to march through a one-year time period and overall energy flows are accumulated on a monthly

1) Solid Wall: No thermocirculation is allowed.

2) Trombe Wall: Thermocirculation is allowed only in the normal direction as previously described. Reverse thermocirculation, as would normally occur at night, is not permitted. (This can be implemented with thin plastic film passive damper draped over the inside of the top opening.) It was determined that if reverse thermocirculation is not prohibited, then the vents are a net thermal disadvantage to the building.

3) Trombe Wall with Control: The result of the normal thermocirculation frequently is to overheat the building during the day. In this option, the vents are closed whenever the building temperature is 75 F or greater. This greatly reduces the required cooling. This presumably would require some passive or active mechanism.

Another configuration is the "water wall." This might consist of cans or drums of water stacked to form a thermal storage wall. Alternatively, vertical freestanding cylindrical tubes could be used or any other means of containing water in a thermal storage wall. When heated by the sun on one side, the water will freely convect to transport the heat across the wall horizontally and thus temperature gradients across the wall will be very small. The water wall has been analyzed by replacing Nodes 6,7,8,9,10 and 11 in Figure 11 with a single Node 6 which represents all of the water mass.

In the analysis the room temperature (Node 2) is always maintained within bounds, T_{min} and T_{max}, which are set at 65 F and 75 F respectively. In the solution of the equations, two possible situations can result:

1) The calculated room temperature falls within the prescribed bounds T_{min} and T_{max}.
2) The calculated room temperature is above T_{max} or below T_{min}.

In the first situation, the room temperature assumes the calculated value. The second situation results in two further possibilities:

1) The calculated room temperature is less than T_{min}. In

this instance auxiliary energy is calculated as required to hold the room at T_{min}.

2) The calculated room temperature is greater than T_{max}. In this instance excess heat is dumped (to the environment, presumably by ventilation) in order to hold the room at T_{max}.

The form of the solution is different depending on whether the room temperature is constrained to a bound or floating in between. In the cases where a transition between operating modes occurs during the hour, the hour is partitioned and the appropriate equations used in each segment.

COMPARISON WITH TEST ROOM RESULTS

The validity of the simulation analysis techniques for two simple concepts is established by demonstrating that they adequately predict the temperature behavior of several small passive test rooms located in Los Alamos, New Mexico. These test rooms are small insulated structures measuring five-feet wide by eight-feet deep by ten-feet high. The south five-foot by ten-foot exposure is glazed with two sheets of Plexiglas. One test room contains four twelve-inch diameter fiberglass tubes which stand eight-feet high directly behind the glass. The tubes are blackened to absorb solar radiation and filled with 188 gallons of water. A second test room contains a 16-inch concrete wall made of stacked cast blocks. Vents at the top and bottom can be opened to allow natural thermocirculation of air. The test rooms are virtually massless except for the thermal storage walls.

On a typical mid-winter clear day (January 11) with an average ambient temperature of 23 F the water temperature at mid-height varied from 82 F to 101 F and the room interior globe temperature varied from 74 F to 94 F. Stratification of water temperature of up to 32 F was observed from the bottom of the tube to the top of the tube. In the second test cell, the exterior wall surface temperature varied from 93 F to 153 F the interior wall surface temperature varied from 84 F to 96 F, the top vent temperature varied from 84 F to 132 F and the room interior globe temperature varied from 74 F to 98 F. During the longest stormy period of the

Solar Architecture

winter, for which the ambient temperature held at roughly 20 F the interior temperature of both cells dropped to the yearly minimum of 48 F.

The thermal behavior of both cells is simulated mathematically using a thermal network of 6 to 11 interconnected points, each of which represents a characteristic location within or outside the test cell such as air temperature, surface temperature, glazing temperature or the temperature of the thermal storage material at some depth within the material. Energy balance equations are written for each location and solved for the hourly observed ambient temperature, solar radiation and wind conditions. The performance of both test cells can be predicted within roughly ± 2 F under most conditions and about 8 F at the extremes.

An additional 12 test rooms have recently been completed for evaluating various passive concepts. Simulation analysis of these will also be carried out for comparison with observed temperature histories.

Test Room Construction
A plan view of the pair of test rooms is given in Figure 12. The walls, floor and ceiling are all 2x4 stud-wall construction with 3 1/2-inch fiberglass insulation. The exterior is covered with plywood sheeting and the interior is lined completely with one-inch polystyrene insulation and all seams are well caulked. Thus the building interior mass is negligible compared with that of the various thermal storage elements added for the various thermal storage elements added for the tests. The net calculated thermal conduction coefficient (exclusive of the south wall and the common wall) between the building interior and the ambient exterior temperature is 0.23 Btuh/F/ft^2 of south glazing.

The entire 50-square foot south wall of each test room is glazed with two sheets of 1/8-inch Plexiglas sheet separated by a 1/2-inch air gap. The rooms are actually oriented with the south wall facing 13° east of due south.

Storage Walls

Two different thermal storage wall concepts have been evaluated in the test rooms. The "water wall" consists of four 12 diameter fiberglass tubes which are free-standing behind the glazing separated by 2.4" from one another and the side walls. During most of the winter, these spaces and the two-foot space above the tubes were blocked with 2"-thick polystyrene insulation cut to fit.

The second test room contained a thermal storage wall. An original wall consisting of 8" solid blocks of cinder block was replaced for most of the winter with a 16" wall stacked from 6" x 8" x 16" blocks of solid cast concrete. Three 3" x 8" holes were left open near the bottom and also near the top to allow air to thermo-circulate from the room floor level up through the 6" space between the wall and the glazing and return at the room ceiling level. The holes were blocked during portions of the test year, left open day and night during other times, and opened during the day only at other times. Storage masses were as follows:

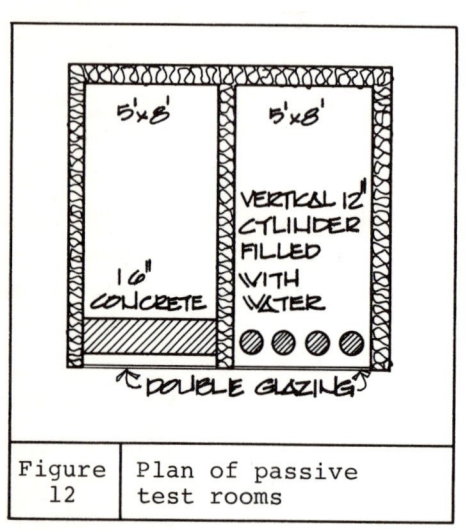

	Thermal Storage/Glazed Area Btu/F-ft^2_g
Water-wall	35
Masonry wall	32.5

Figure 12: Plan of passive test rooms

General Results

The test rooms were very well heated by the sun. Average room temperatures were 60 to 70 F above the ambient average temperature during typical sunny midwinter days. The minimum temperature in the one room was very nearly equal to the minimum temperature in the other room each night throughout the winter.

Solar Architecture

Daily room temperature variations were approximately as follows:
- Water wall: 20 F
- Masonry wall:
  - Vents closed   11 F
  - Vents open   24 F

Thus the main effect of the thermocirculation vents is to provide direct heat to the room during the day. This increases the daily swing and although the daily average room temperature is larger by about 5 F, it does not noticeably change the minimum room temperature. The maximum inside wall surface temperature occurs at roughly 4:00 p.m. on the water wall and at 8:00 to 10:00 p.m. on the masonry wall.

Simulation Model

Thermal network models of the two test-room configurations are shown in Figure 11. The thermal conductances used in the analysis are as follows (all values are normalized to one square foot of glazing, ft_g^2): Water wall and masonry wall:

U_1 = "load" = 0.23 Btu/hr F ft_g^2
U_2 = radiation + exterior film conduction

$$\text{radiation} = \frac{(T_1^2 + T_3^2)(T_1 + T_3)}{1/\varepsilon_1 + 1/\varepsilon_3 - 1}$$

exterior film condution = 4 Btu/hr F ft_g^2

(for an average wind speed of 7.5 mph)

α = Stefan-Boltzmann Constant = 1.713 x 10^{-9} Btu/hr ft^2 F^4
ε_1 = 0.8 (Plexiglas)
ε_3 = 0.8 (ambient)
$U_3 = U_6 = U_7$ = 1.0 Btu/hr F ft_g^2
U_4 = radiation + conduction
   radiation similar to U_2 with $\varepsilon_4 = \varepsilon_5$ = 0.8
   conduction = 0.36 Btu/hr F ft_g^2, (1/2" space)
U_5 = radiation similar to U_2 with $\varepsilon_5 = \varepsilon_6$ = 0.8

Masonry Wall:

$U_8 = U_{12}$ = 4.7 Btu/F ft_g^2 (2" concrete)
$U_9 = U_{10} = U_{11}$ = 2.35 Btu/F hr ft_g^2 (4" concrete)

The mass heat capacity of all nodes is zero except as follows:

Water Wall: Node 6 heat capacity = 35 Btu/F ft_g^2
Masonry Wall: Nodes 7, 8, 9 and 10 heat capacity = 8.1 Btu/F ft_g^2

Input to the simulation model are instantaneous hourly values of measured ambient temperature (Node 1) and also the solar radiation measured on a vertical surface parallel to the glazing integrated over one hour.

Solar transmittance through the glazing is calculated accounting for both Fresnel reflections and for absorption in the glazing based on the calculated incidence angle. This leads to an energy source term into Node 6. For the water wall, 30% of the transmitted solar radiation is reflected out by the light polystyrene blocks between and above the tubes. For the masonry wall this is 20%.

When the vents are open the air flow rate due to thermocirculation is estimated assuming the major flow resistance to be in the vents. The temperature distribution in the wall-glass air space is assumed to be linear so that T_5 is the arithmetic average of the inlet and outlet air space temperatures. The room temperature is assumed to be constant, so the inlet air space temperature is the room temperature, T_2.

When the vents are closed there is no thermocirculation accounted for and all the solar heat transfer to the room is by conduction through the wall.

Temperatures are calculated each hour as required to achieve an energy balance at each node point. Since the thermal conductances are nonlinear, this is done by iteration around a linear equation solving routine.

Comparison
Comparison of calculated and measured temperatures have been made for the period December 31, 1976 to January 6, 1977 during which the thermocirculation vents were open and during the

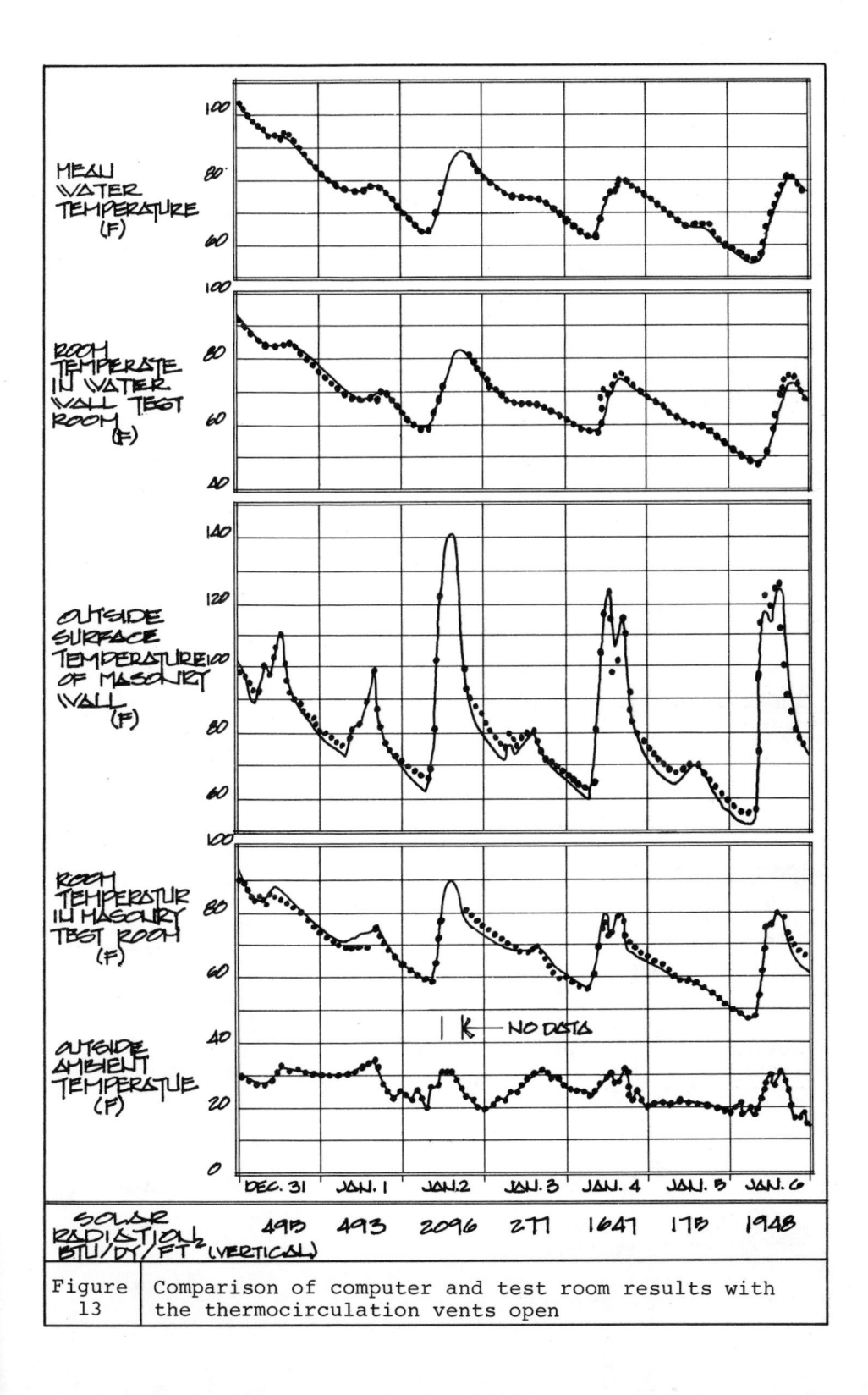

Figure 13: Comparison of computer and test room results with the thermocirculation vents open

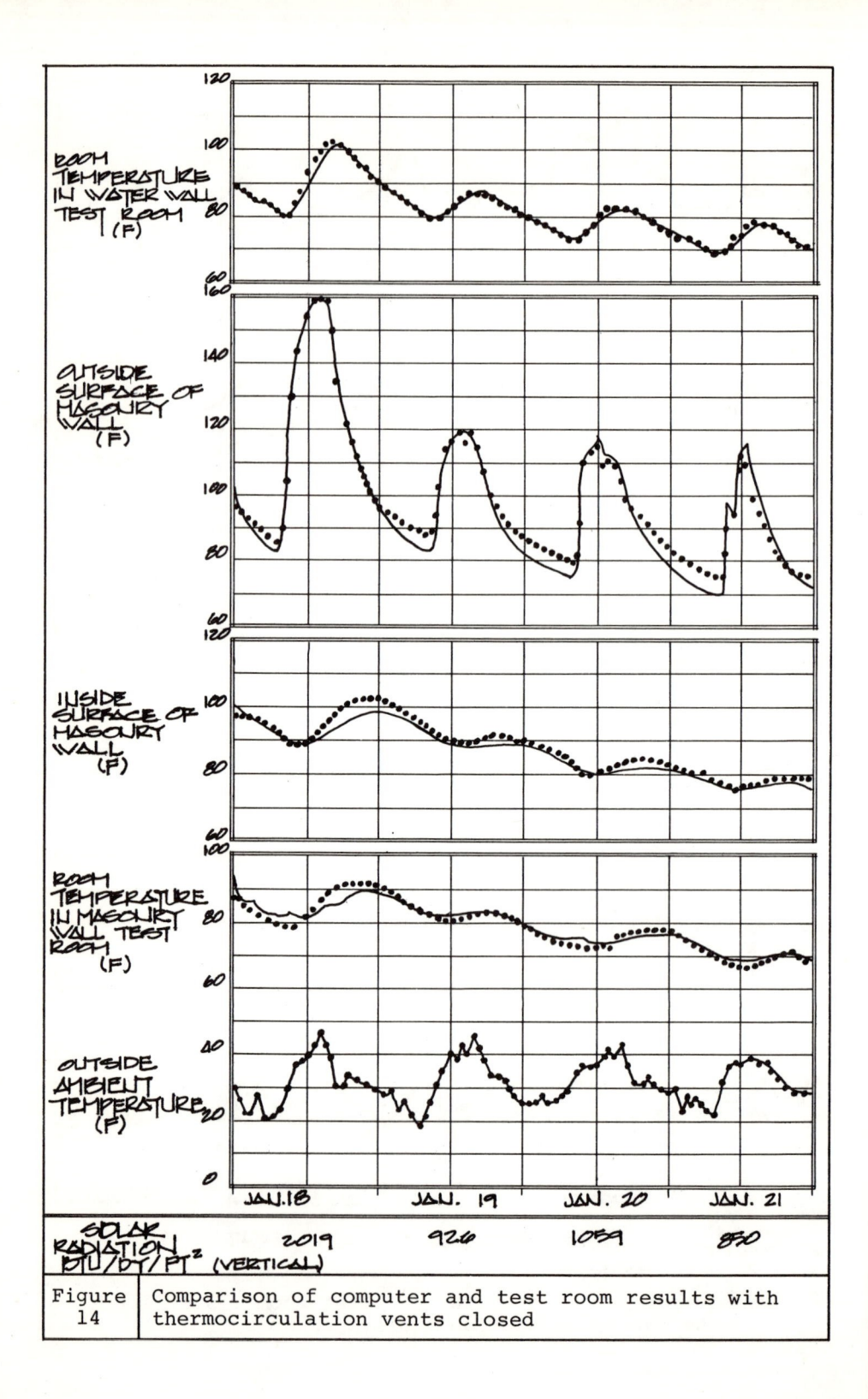

Figure 14. Comparison of computer and test room results with thermocirculation vents closed

Solar Architecture

period January 18 to 21 when the vents were blocked. The first interval is a period of erratic weather with three separate snowfalls and some bright sunny intervals. The second period starts with a sunny day followed by three days of partial sun.

The simulation model quite accurately predicts the temperature during most of the time--generally within ±2 F. The largest errors occur during strong heating periods when discrepancies up to 8 F are observed.

The comparisons are shown on Figures 13 and 14. For each graph, the symbols • represent the measured quantity and the solid line represents the value calculated by the simulation model.

It is concluded that these particular passive collector-storage elements are quite amenable to accurate representations using the type and level of simulation models employed.

SIMULATION ANALYSIS FOR LOS ALAMOS

For much of the preliminary analysis done to date, the solar and weather data used were for the Los Alamos year September 1972 to August 1973. For this year, the total radiation on a horizontal surface was 518,000 Btu/ft^2 and the space heating load (based: 65 F) was 7350 degree-days (18% higher than normal for Los Alamos). This is a severe test. For these initial studies the glass conductance was characterized by a single constant term rather than the non-linear, two-term representation shown in Figure 11. The results are useful to determine general effects but the simple model over-predicts the total performance by about 12%. Five different cases have been studied as follows: (see Figure 15 for designation of symbols).

Case 0: The room and storage are the same temperature (U_1 is infinite.)

Case 1: Storage is coupled thermally only to the room. This case would represent massive internal walls or furniture placed out of the direct sunlight ($U_{gs} = 0$, $U_{ws} = 0$).

Case 2: Storage is placed directly in front of the glass. The sun shines on and is absorbed by storage. Storage is thermally coupled to the environment through the glass and also to the room. ($U_{gr}= 0$, $U_{ws}= 0$).

Case 3: Storage is placed against the back wall out of the direct sun. This case would represent massive walls or roof insulated on the outside ($U_{gs}= 0$, $U_{wr}= 0$).

Case 4: Storage is placed in the room in the direct sun but loses heat only to the room. ($U_{gs}= 0$, $U_{ws}= 0$).

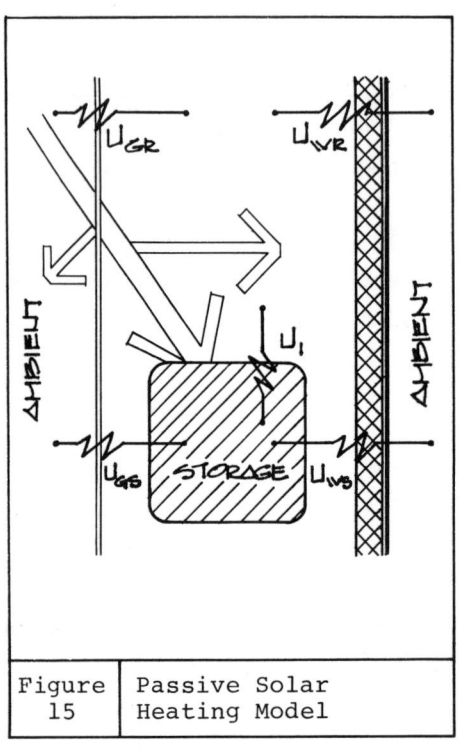

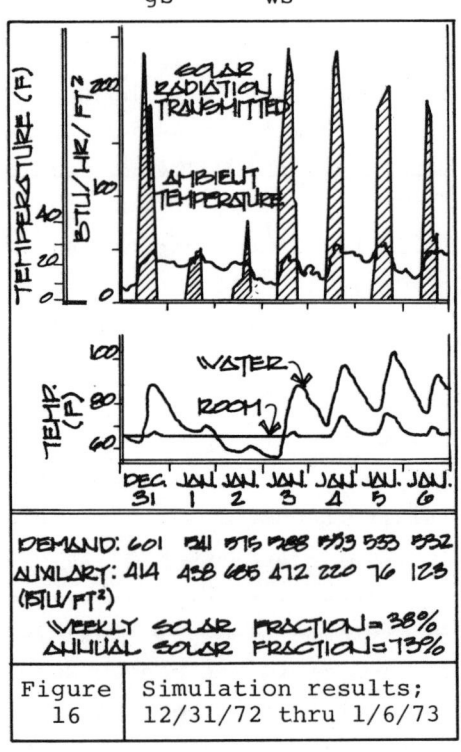

Figure 15	Passive Solar Heating Model

Figure 16	Simulation results; 12/31/72 thru 1/6/73

These cases are intended to represent extremes. Any real design may tend toward one or another case but will actually be a mixture. For the preliminary study the U values were held constant although in reality, they will vary with temperature difference, wind, and other influences. The glass was assumed to be vertical, to face due south, and to be unshaded. Three glazings were studied:

1) Single glazing: $U_g = 1.1$ Btu/hr·ft^2 - F.
2) Double glazing: $U_g = 0.50$ Btu/hr·ft^2 - F.
3) Night insulated double glazing: $U_g = 0.1$, 5 pm to 8 am

Solar Architecture

A value of U_{wr} of 0.5 Btu/hr·ft$_g^2$ - F was chosen initially.

A storage mass of 30 Btu/ft$_g^2$ - F was chosen. This is equivalent to 30 pounds of water per sq ft of glass or 150 pounds of concrete per sq ft of glass.

Lastly the room temperature was allowed to vary 5 F around a desired value of 70 F. Therefore T_{min} = 65 F and T_{max} = 75 F.

Simulation Results for a Case with Isothermal Storage

It is instructive to observe the simulation results for a few days of cold weather. Figure 16 shows the 7-day interval between the 31st of December and the 6th of January. Case 2 was chosen for this calculation with a value of U_3 = 1.0 Btu/hr-ft$_g^2$ - F. There was snow on New Year's Eve followed by two days of cloudy weather and then cold but sunny weather.

In the following discussion the effect of variations in various

Month	Building Load	Incident Solar Energy On Glass Wall (Btu/ft2)	Solar Energy Transmitted Through Glass Wall (Btu/ft2)	Excess Energy Ventilated (Btu/ft2)	Auxiliary Energy Required (Btu/ft2)	Degree days (F) (65 F Base)	Percent of Solar Heating
Sep	6073	36882	25311	6183	48	208	99.20
Oct	8447	31623	22044	3479	2149	506	74.56
Nov	13617	40108	29331	638	4300	1032	68.42
Dec	15006	39693	29638	762	5788	1144	61.43
Jan	15865	46150	34353	817	4573	1202	71.17
Feb	13066	38830	27956	961	4053	982	68.98
Mar	12650	34802	23904	472	5377	980	57.49
Apr	10451	34288	22682	650	3081	759	70.52
May	6861	30510	19856	2583	842	366	87.72
Jun	4147	33966	22162	8321	1	123	99.96
Jul	3520	28398	18686	7413	0	28	100.00
Aug	3252	31298	20660	8873	0	16	100.00
Annual	112964	426547	296582	41152	30214	7350	73.25
Figure 17	Case 2, Solar Performance Summary						

parameters will be shown. In each case the simulation model was run repeatedly varying one parameter at a time while holding the others constant at the nominal values given above. The circled point on each graph represents the nominal case.

Figure 18 shows the yearly results for all five cases as a function of the room-to-storage thermal coupling factor, U_I. All cases become equivalent for large values of U_I. Cases 2 and 4 are observed to be appreciably better in performance than the others. These two cases are for a wall directly heated by the sun. Clearly this is an enormous advantage. Case 4 is unrealistic for low values of U_I because the sun must shine through the room to reach storage and this implies transparent insulation. Case 2 is fairly representative of a "drum-wall," in which the thermal storage is in water contained in cans placed in front of the glass wall.

It is significant to note that there exists an optimum value of U_I for Case 2. The optimum value is approximately 1.5 Btu/hr·F-ft$_g^2$. The reason for the optimum is as follows: At higher values of U_I, storage loses too much heat to the room during charging periods. This prevents storage from attaining higher temperatures and storing greater amounts of heat. At lower values of U_I, storage attains such high temperatures during

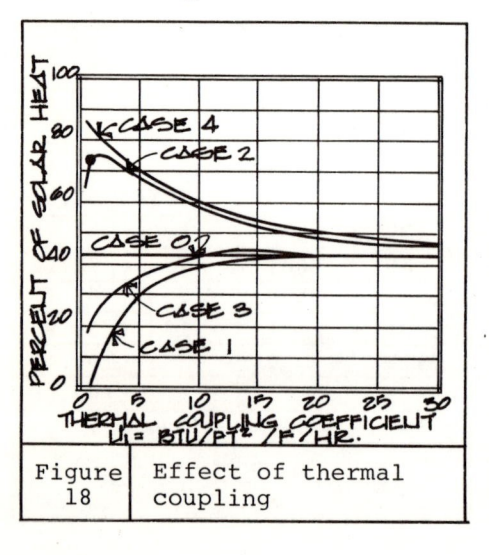

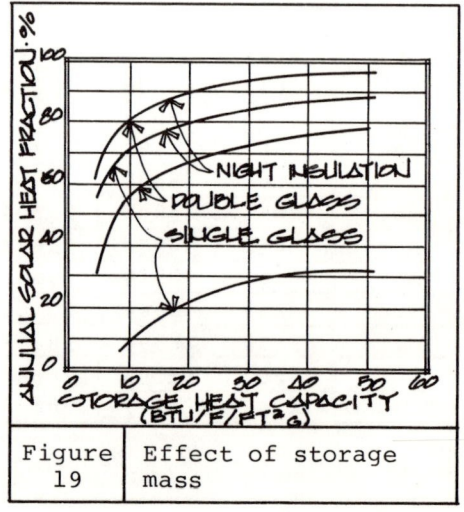

Figure 18	Effect of thermal coupling

Figure 19	Effect of storage mass

Solar Architecture

charging periods that it loses too much heat through the glass to the environment. Clearly the optimum value of U_I will depend on the chosen storage heat capacity.

The effect of varying storage heat capacity and glass insulation is shown in Figure 19. Most of the benefits of storage are obtained at a value of 30 Btu/F-ft^2_g. The improvement obtained with double glazing is very dramatic. In fact, a single-glazed wall without night insulation can hardly be considered a viable passive solar heating element since only 30% solar heating can be achieved even with large storage and the glass is a net loser at low storage. The increased effectiveness of insulating the glass at night (for example, as in a beadwall) is impressive. The cost-effectiveness of this approach needs further study. Night insulation can be seen to be far more important with single glazing than with double glazing. Single glazing becomes viable only with night insulation. A strategy of placing night insulation based on observed conditions rather than a timeclock would result in only a small increase in performance (2%).

The effect of varying the glass area is inverse to the effect of varying the building thermal load U_w. This is shown in Figure 20.

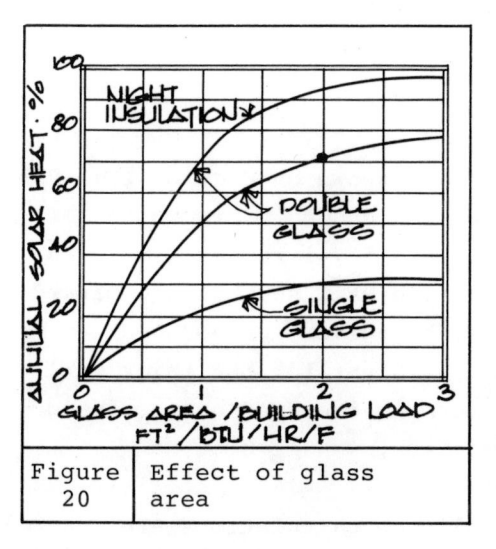

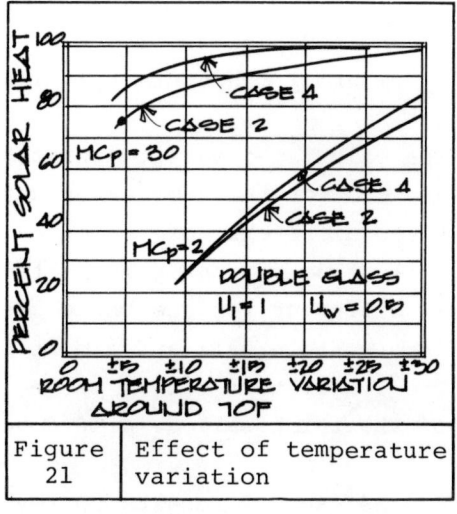

| Figure 20 | Effect of glass area | Figure 21 | Effect of temperature variation |

The effect of the allowable temperature swing is shown in Figure 21. A variation of ±5 F may be reasonable for a residence, whereas a much larger variation may be tolerable in other buildings such as a warehouse or greenhouse. The effect of mass is also shown for both Cases 2 and 4.

Simulation with a masonry wall

The example of Case 2 is somewhat representative of the concept developed at Odeillo, France by Trombe and his colleagues utilizing a masonry wall, specifically a concrete wall. In the French concept, a thermocirculation path was provided by perforations extending through the wall at the top and bottom. The value of these perforations had not yet been established and this effect was not simulated.

In order to study the basic performance characteristics of such a wall, as compared to the case of an isothermal wall studied earlier, the mathematical model was modified to describe the time and one-dimensional space dependent thermal transport of heat through the wall. This was done by simulation of the masonry temperature at the wall surfaces and at several different distances into the wall.

The thermal properties used for the masonry were as follows (typical of dense concrete):
    Heat Capacity: 30 Btu/ft^3 F
    Thermal Conductivity: 1 Btu/ft F hr

The calculated wall temperatures are shown on Figures 22a, 22b, and 22c for the same seven-day period shown in Figure 16 for three different wall thicknesses--0.5 ft, 1 ft and 2 ft. The daily fluctuations felt on the inside wall surface are markedly different for the three cases, being very pronounced (45 F) for the thin wall and almost non-existent for the thick wall. The longer-term effect of the storm is observed on the inside of the thick wall as a 10 F variation.

The net annual results of several such calculations are

Solar Architecture

summarized in Figure 23. The net annual thermal contribution of the three different thicknesses of walls are not markedly different. In fact, the one foot thick wall is the best of the three--giving an annual solar heating contribution of 68%. This compares with a value of 73% for an isothermal wall with the same heat capacity. In each case auxiliary cooling or heating was assumed to maintain the room temperature within the bounds given previously. Although the net thermal contribution of the thin wall and thick wall cases are nearly the same, the amount of control required for the thick wall is much less and the

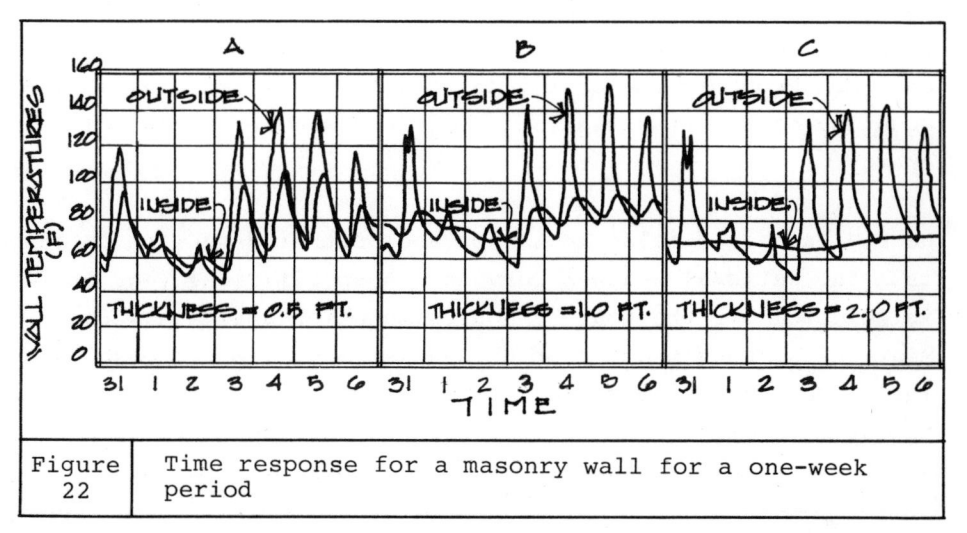

| Figure 22 | Time response for a masonry wall for a one-week period |

variation in room temperature within the set bounds is much less.

The effect of variations in wall thermal conductivity is also shown in Figure 23. The isothermal wall corresponds to the infinite conductivity case. It can be seen that for each conductivity there exists a thickness which will give a maximum yearly solar energy yield. The optimum thickness

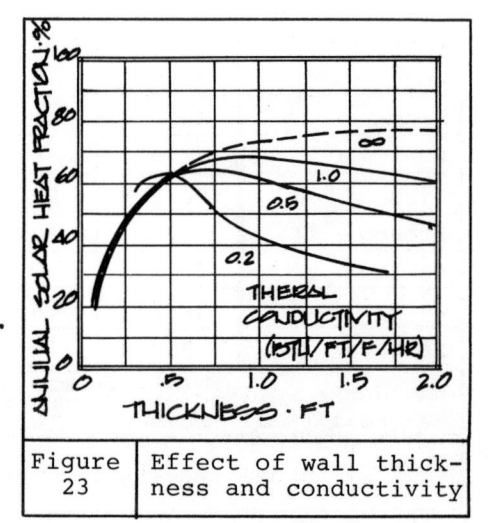

| Figure 23 | Effect of wall thickness and conductivity |

decreases as the thermal conductivity decreases. Annual heating performance, of course, is only one consideration in the selection of wall materials and wall thicknesses.

Results for Other Climates

In the analysis for other climates, the more detailed simulation model of the glazing layers was used as shown in Figure 11. This more accurate representation makes a significant difference--the annual solar heating fraction for Los Alamos was reduced from 68% to 56% in one case. Since the more complex model has been validated against the test room data, it is more reasonable. Several changes in the model are responsible for the change. Probably the most significant is a more detailed accounting of the transmission of diffuse and reflected solar energy.

Hourly values of solar radiation and weather data were obtained from the National Weather Service. A specific one-year period is selected for the hour-by-hour simulation analysis. For Madison, Wisconsin the year chosen is July, 1961 through June, 1962. The results of a parametric study of the effect of thermal storage mass is shown in Figure 24 for the four cases

City	Annual Percent Solar Heating				
	WW	SW	TW	TW(A)	TW(B)
Santa Maria	99.0	98.0	97.9	97.3	98.0
Dodge City	77.6	69.1	71.8	62.8	73.6
Bismarck	49.8	41.3	46.4	31.1	47.6
Boston	60.0	49.8	56.8	44.9	56.7
Albuquerque	90.8	84.4	84.1	81.8	87.5
Fresno	85.5	82.4	83.3	78.0	83.4
Madison	43.1	35.2	41.6	24.7	42.0
Nashville	68.2	60.7	65.2	54.1	65.4
Medford	59.0	53.3	56.1	42.2	56.8
WW: Water wall TW(A): Trombe wall w/vents open at all times TW(B): Trombe wall w/thermostatic vent control SW: Solid wall (no vents) TW: Trombe wall (no rev. vent flow)					
Table I	Annual results for thermal storage				

Case: 18 in. Trombe Wall; No reverse thermocirculation
Thermal conductivity = 1 Btu/ft/hr F
Heat Capacity = 30 Btu/ft^3 F
Vent Size = 0.074 ft^2/ft of length (each vent)
Load (U_l) = 0.5 Btu/ft^2 F hr
Temperature band = 65 F to 75 F

City	Year Starting	Heating Degree-Days	Latitude	Solar Heating Btu/ft^2	Solar Heating Fraction, Percent
Los Alamos, NM	9/1/72	7350	35.8	60,200	56.5
El Paso, TX	7/1/54	2496	31.8	50,000	97.5
Ft. Worth, TX	7/1/60	2467	32.8	38,200	80.8
Madison, WI	7/1/61	7838	43.0	44,900	41.6
Albuquerque, NM	7/1/62	4253	35.0	63,600	84.1
Phoenix, AZ	7/1/62	1278	35.5	38,300	99.0
Lake Charles, LA	7/1/57	1694	30.1	34,300	90.5
Fresno, CA	7/1/57	2622	36.8	43,200	83.3
Medford, OR	7/1/61	5275	42.3	47,400	56.1
Bismarck, ND	7/1/54	8238	46.8	53,900	46.4
New York, NY	6/1/58	5254	40.6	48,000	60.2
Tallahassee, FL	7/1/59	1788	30.3	40,700	97.3
Dodge City, KS	7/1/55	5199	37.8	58,900	71.8
Nashville, TN	7/1/55	3805	36.1	39,500	65.2
Santa Maria, CA	7/1/56	3065	34.8	69,800	97.9
Boston, MA	7/1/57	5535	42.3	47,100	56.8
Charleston, SC	7/1/63	2279	32.8	47,900	89.3
Los Angeles, CA	7/1/63	1700	34.0	53,700	99.9
Seattle, WA	7/1/63	5204	47.5	42,400	52.2
Lincoln, NE	7/1/58	5995	40.8	53,500	59.1
Boulder, CO	1/1/56	5671	40.0	62,500	70.0
Vancouver, BC	1/1/70	5904	49.1	46,000	52.7
Edmonton, ALB	1/1/70	11679	53.5	37,700	24.7
Winnipeg, MAN	1/1/70	11490	49.8	33,700	22.6
Ottawa, ONT	1/1/70	8838	45.3	37,900	31.9
Fredericton, NB	1/1/70	8834	45.8	40,100	33.9
Hamburg, Germany	1/1/73	6512	53.2	24,900	27.5
Denmark	?	6843	56.0	43,100	43.8
Tokyo, Japan	?	3287	34.6	50,300	85.8

Table II Annual solar heating results for twenty-nine various climates

studied. As had been noted in a previous preliminary analysis, there is an optimum thickness of about one foot for the masonry wall. From calculations done for other locations it is determined that this optimum does not depend on climate.

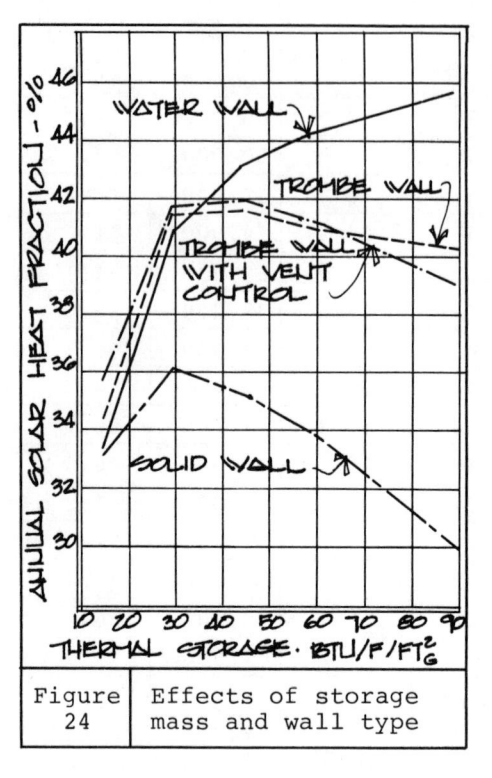

A study of the effects of climate on performances is given in Table I. These calculations are all for a thermal storage mass of 45 Btu/F ft$_g^2$ (18" of concrete or 8.6" of water). Although some cases are clearly better than others, all seem to be viable approaches to solar heating in all the climates studied. The effectiveness of the thermocirculation vents is pronounced in the colder climates.

| Figure 24 | Effects of storage mass and wall type |

The ultimate measure of cost-effectiveness of these concepts will be the heating energy delivered to the building by the solar wall. These annual values are given in Table II for the particular case of the 18" Trombe wall.

BIOGRAPHICAL SKETCH

J. Douglas Balcomb is the assistant Division Leader for Solar at the University of California, Los Alamos Scientific Laboratory in Los Alamos, New Mexico. Dr. Balcomb is also the Chairman of the New Mexico Solar Energy Association. He has developed computer programs for passive solar designs.

ENDNOTES

1. M.G. Davies, "The Contribution of Solar Gain to Space Heating," Paper 47-1 in Extended Abstracts of the 1975 International Solar Energy Congress and Exposition, UCLA, Los

PREDICTING THE PERFORMANCE OF PASSIVE SOLAR HEATED BUILDINGS

Presented by: Ed Mazria
Prepared by: Ed Mazria, M.S. Baker, F.C. Wessling

Rising energy costs have provided the incentive to reduce energy consumption through the design and application of alternative energy sources. In order to have widespread (global) application, it is necessary that this technology be inexpensive and simple in concept. Passive solar energy systems offer such an alternative.

There are many types of passive solar heating systems being built today, but the simplest and most efficient is commonly referred to as a "direct gain" system. Solar energy admitted into a space through windows, clerestories and/or skylights is absorbed and stored within the walls or floor of the space for use during the evening. The walls and/or floors are constructed of materials capable of storing heat, such as brick, concrete, adobe and water (contained). In effect, the entire space becomes the solar heating system. There are no separate collector panels, storage units or mechanical equipment. By properly designing and integrating all the architectural elements within each space (doors, windows, walls, floor and roof), a passive solar heating system is able to utilize most of the solar radiation admitted into a space for winter heating.

Our present knowledge of passive solar heating is a result of
the determined efforts of a group of designers and builders.
To date, relatively little scientific research support has been
allocated for passive solar heating projects. In the absence
of such research, most of our present information has been
learned through the performance of various existing projects.

Solar gain through south-facing glass is easily calculated.
However, predicting the performance of thermal mass materials
in a space is beyond the capability of most building designers.
An analytical solution has been developed for a simplified model
of passive solar heating systems to establish usable performance
criteria for their design. This model is presently being
used to develop rules-of-thumb and other analytical and graphic
tools for designers of passive solar heated buildings. Our
studies indicate that passive solar heating systems can supply
a significant portion of a building's winter space heating
requirements and maintain relatively stable indoor air temperatures.
The preliminary conclusions presented in this article
for the design of "direct gain" passive systems have considerable
implications for future building design.

ANALYTICAL MODEL

A graphic description of the analytical model used in this
study is represented in Figure 1. Indoor air temperature is
represented by T_i; outdoor air temperature by T_o. Solar energy
enters the space through glazing and strikes the interior face
of a thermal mass wall which has a thickness L. The value Q_s
equals the solar energy incident per unit area of mass wall.
The portion of incident solar energy absorbed by the mass wall
is determined by the surface absorptivity α. Solar energy
reflected from the face of the wall is either lost to the outside
(reflected back out through the glazing) or reflected onto
other non-massive interior surfaces which act as heat source
Q_f to the room air. In this paper, it is assumed that any
reflected sunlight is added as heat directly to the room air.

The thermal storage wall loses heat to the outside by two paths.

Solar Architecture

The first path is through the insulation on the exterior surface of the storage material. The thermal conductance of the insulation is represented by U_o. The second path is heat transfer from the surface of the mass to room air, and from room air through the exterior skin of the building to the outside. This heat transfer depends on the following thermal conductances: U_w, from the wall surface to room air, and U_h, from the room air to the outside.

The model represents temperatures that would occur after several consecutive near-identical days, such as several clear or cloudy days in a row. The results for consecutive clear days are presented here since this has a greater influence on system design.

The analytical model with minor changes can also be used to analyze other passive solar heating systems such as the Trombe wall and solar greenhouse/housing combinations. This is being investigated and will be presented in a future paper.

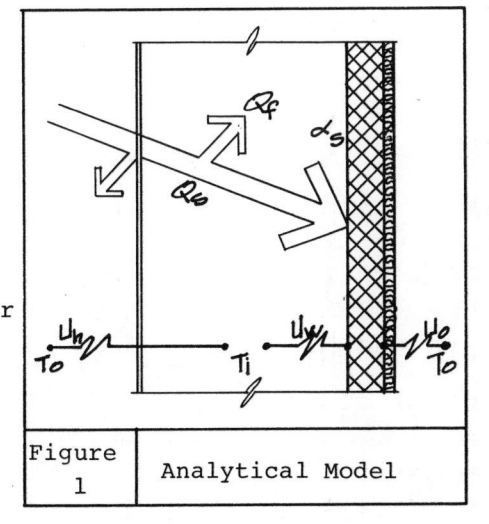

Figure 1. Analytical Model

SYSTEM DESCRIPTION

In a "direct gain" system, because the building configuration used in coupling solar radiation with a thermal mass greatly influences system performance, three different systems are presented for investigation.

System 1: Thermal storage mass is placed against the rear wall in direct sunlight. The surface area of mass exposed to direct sunlight over the day is 1.5 times the area of the glazing. This system represents a space with a horizontal band of south-facing windows or clerestories coupled directly to a thermal mass which is insulated on the exterior face (Figure 2).

System 2: Thermal storage mass is placed against the rear wall in direct sunlight. The surface area of mass exposed to direct sunlight over the day is three times the area of the glazing. This system represents a space with vertical windows (evenly spaced) or translucent (diffusing) glazed openings (Figure 3).

System 3: The space (walls and floor) becomes the thermal storage mass. The surface area of mass exposed to direct sun-

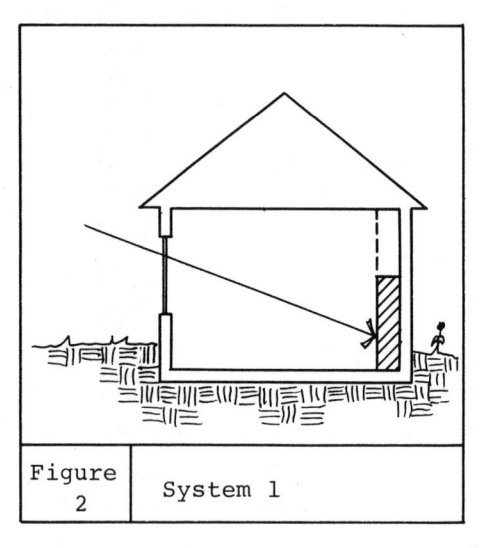

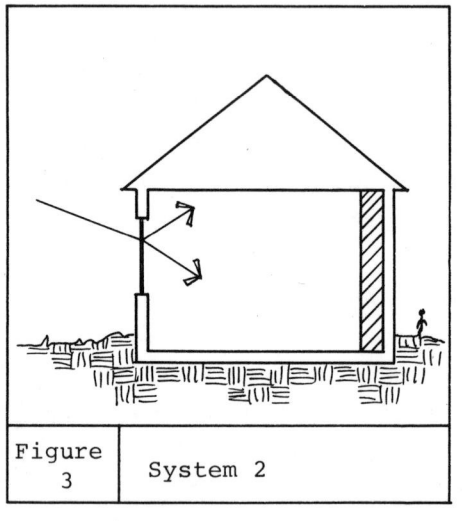

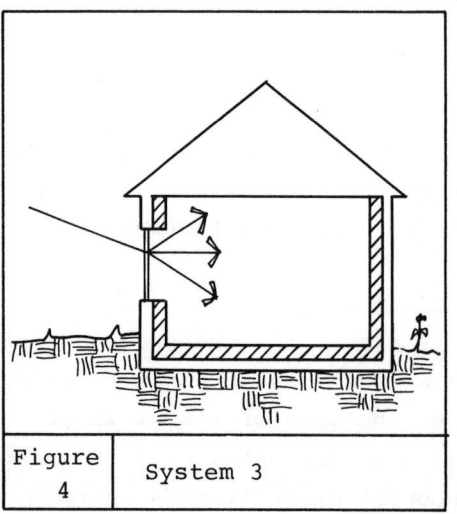

Figure 2 — System 1

Figure 3 — System 2

light is nine times the area of the glazing. This system represents a space constructed of masonry materials with translucent glazed openings or clear glazed openings coupled initially with light-colored interior surfaces (Figure 4).

COMPUTER SIMULATION

Each system is analyzed by computer for different parameters of outdoor air temperatures,

Figure 4 — System 3

Solar Architecture

latitude, thermal storage material, thickness and surface color. The building is considered to be well insulated with a space heat loss (U_h) of 0.35 Btu/hr ft^2 (floor area) F. Heat loss from the thermal storage mass to the exterior is calculated separately (U_o = 0.08 Btu/hr ft^t F) and is additional to the space heat loss. The overall U value of the building (U_h, space heat loss + U_o, storage mass heat loss) is 0.39 Btu/hr ft^2 (floor area) F for System 1, 0.41 Btu/hr ft^2 F for System 2, and 0.53 Btu/hr ft^2 F for System 3.

An analysis using January clear-day insolation values and outdoor temperatures for Portland, Oregon (45°NL) is presented in this paper to illustrate the computer results. All systems assume 75% solar absorption by the thermal mass, with the remaining 25% supplied directly to the air in the space as heat. South-facing glass is 25% of the building's floor area. This percentage is determined by the amount of solar gain needed to offset the building's heat loss for an average clear January day with an indoor design temperature of 70 F. System performance is measured using indoor air temperatures. In reality, system performance and human comfort are determined by a complex set of interior conditions (air temperature, mean-radiant temperature, relative humdity and air movement).

SIMULATION RESULTS: Thermal Storage Thickness

Design Parameters
Systems 1, 2 and 3 are analyzed for a concrete thermal storage mass of different thickness.

System 1: During a clear day, an increase in mass thickness beyond 8" results in little improvement in system performance. Figure 5 illustrates the indoor air temperatures (over a day) for mass thickness of 4", 8" and 16". By increasing the mass (thickness) from 4" to 8", maximum air temperatures are relatively unchanged, while minimum air temperatures are changed significantly; the 8" mass wall increases the minimum room air temperature 5 F. Increasing mass thickness to 16" has little

Figure 5. Computer Model: System 1

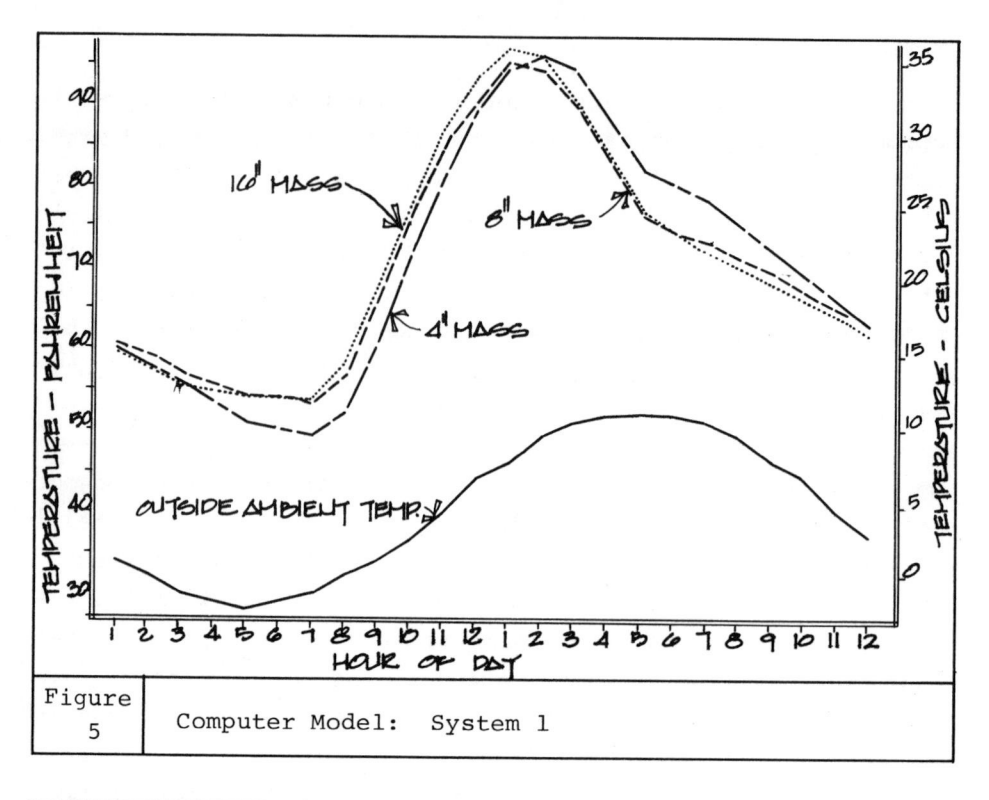

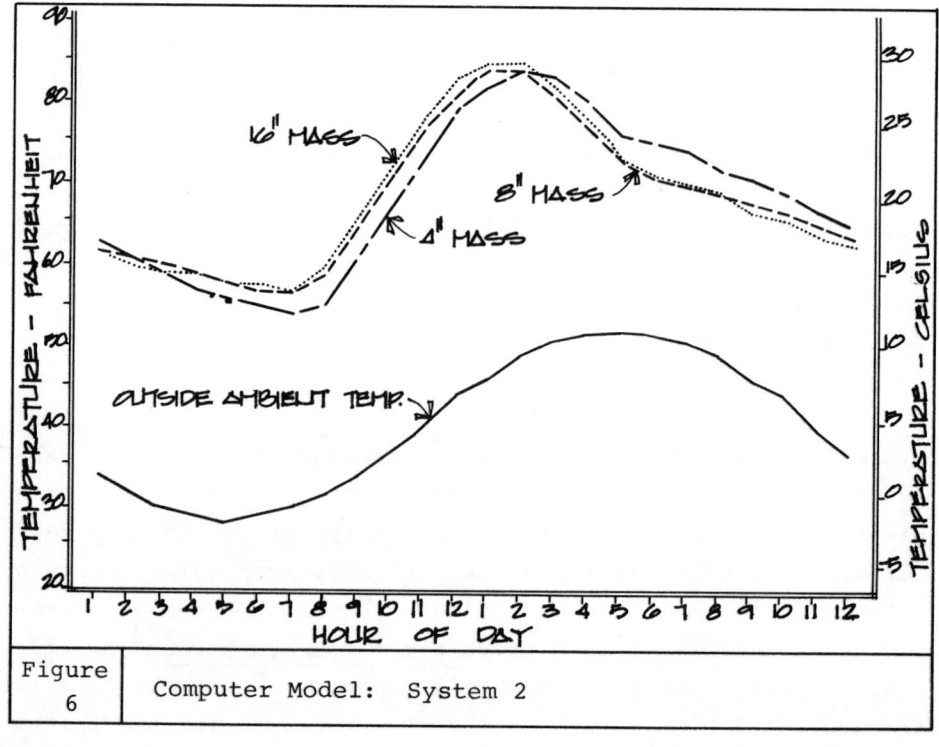

Figure 6. Computer Model: System 2

Solar Architecture

impact on air temperatures. In fact, maximum temperatures in the space are increased slightly (poorer performance) with mass thickness greater than 8".

System 2: An increase in mass thickness beyond 8" results in little change in system performance. Figure 6 illustrates room air temperatures for a mass wall thickness of 4", 8" and 16". The major temperature difference occurs by increasing mass thickness from 4" to 8"; maximum room air temperature remains unchanged while minimum air temperature is raised 2 F. Beyond an 8" thickness, there is very little variation in room temperature.

System 3: An increase in mass thickness beyond 4" results in little change in system performance. For all three thicknesses, room air temperatures are very similar.

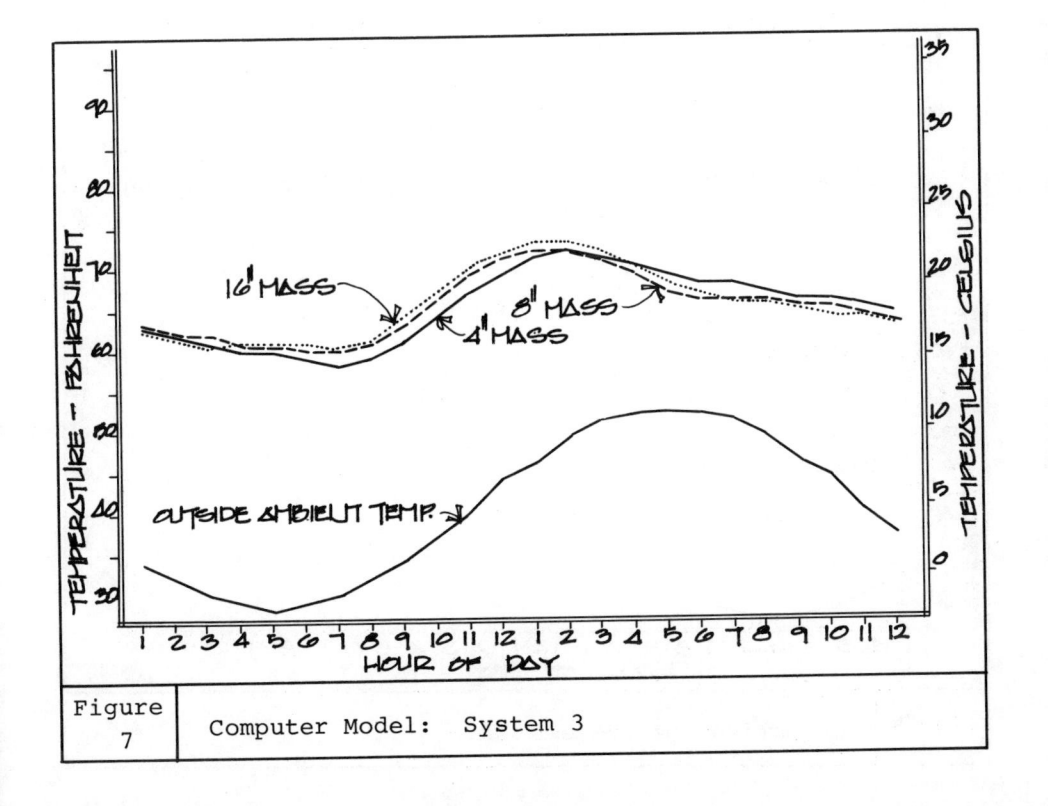

Figure 7. Computer Model: System 3

COMPARISON OF SYSTEMS 1, 2 and 3

By spreading solar radiation over a larger mass surface, room air temperature fluctuations are decreased (see Figure 8 and Table 1). For residential use, where relatively stable indoor air temperatures are desired, System 3 is preferable. System 3 maintains the highest minimum air temperatures and smallest daily temperature fluctuations (±6.5 F). Both System 1 and 2 would require some ventilation to prevent overheating during the day. Ventilation lowers system performance by disposing of excess heat which could be utilized for space heating during evening hours. System 1, with a maximum room air temperature of 95 F, requires a significant amount of daytime ventilation.

By spreading solar radiation over a larger mass surface area, a greater percentage of solar heat is stored in the mass at the end of the day, at 5 pm (see Table 1). System 3 stores the largest percentage of solar heat admitted into the space (61%). The efficiency of stored solar energy in the mass wall (including a glass transmissivity of 75% over the day) is 46%. This does

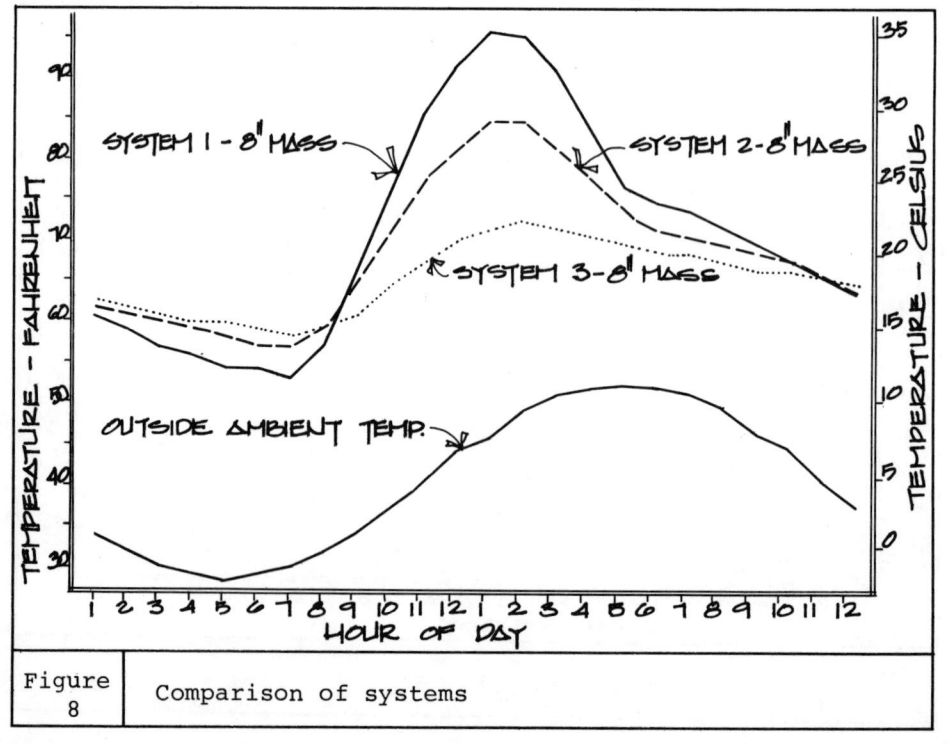

Figure 8 — Comparison of systems

not include heating provided during the daytime hours (7 am to 5 pm). The efficiency of System 3 with this taken into account becomes 64% of the solar radiation incident on the glass (the energy used to maintain indoor temperatures above 70 F is discounted). As can be seen from Figure 8, the highest and lower indoor air temperatures occur at approximately the same time. The lows occur at 7 am, one hour before sunrise and the highs at 1 pm, except for System 3 which peaks one hour later at 2 pm.

The greatest difference between the surface temperature of the thermal mass and indoor air temperature occurs during the late evening and early morning hours. In early morning, even though the indoor air temperature is low, the mean radiant temperature of the space is higher, providing greater environmental comfort than registered by the air temperature alone.

SIMULATION RESULTS: Thermal Storage Materials

Design Parameters
System 2 is analyzed for different thermal storage materials of 8" thickness. These materials include concrete (stone), brick (common), brick (magnesium additive), adobe and water contained in cans (represented as an isothermal mass). These materials have the physical properties listed in Table 2.

RESULTS
The major impact on system performance occurs with a change in the conductivity of the material. By using a thermal mass of higher conductivity, the air temperature fluctuations in the space are minimized. The largest temperature fluctuations occur using adobe, which has the poorest conductivity (maximum 93 F and minimum 53 F). The best performance (smallest temperature fluctuations) occur with brick which has a magnesium additive (maximum 81 F and minimum 57.5 F). When using masonry materials, higher conductivity produces substantially reduced maximum and slightly increased minimum space air temerpatures. A comparison of masonry with water storage (same thickness and modeled as an isothermal wall) shows that for System 2, water storage produces

	System 1 8" thickness	System 2 8" thickness	System 3 4" thickness
Maximum space air temperature	95 F (35°C)	84 F (29°C)	71 F (22°C)
Minimum space air temperature	53 F (12°C)	57 F (14°C)	59 F (15°C)
Space air temperature fluctuation	41 F (23°C)	26 F (14°C)	13 F (7°C)
Thermal mass (average) temperature at 7 am	70 F (22°C)	67 F (19°C)	61 F (16°C)
Thermal mass (average) temperature at 5 pm	90 F (32°C)	78 F (26°C)	70 F (21°C)
Percentage solar stored (at 5 pm, sunset)*	49%	55%	61%

*percentage of solar radiation admitted into the space

Table I — Comparison of systems

	Conductivity (k) Btu/hr ft^2 F (W/m^2 °C)	Specific Heat (C) Btu/lb F (kJ/kg °C)	Density (ρ) lb/ft^3 (kg/m^3)
Concrete (stone)	12.0 (1.70)	0.20 (0.84)	140.0 (2,240)
Brick (common)	5.0 (0.72)	0.20 (0.84)	120.0 (1,920)
Brick (magnesium add.)	26.4 (3.80)	0.20 (0.84)	120.0 (1,920)
Adobe	3.6 (0.52)	0.24 (1.00)	106.0 (1,700)
Water (isothermal)	—	1.00 (4.19)	62.4 (1,000)

Table II — Thermal storage material properties

Solar Architecture

produces the best results or smallest space air temperature fluctuations (maximum 75 F and minimum 62 F). It should be noted that System 3 with 4" of concrete mass achieves similar results as System 2 with an 8" water wall. This analysis does not take into account sizing for cloudy-day storage which will affect mass thickness and area of glazing.

CONCLUSION

Our studies indicated that passive solar heating systems can supply a significant portion of a building's winter space heating requirements <u>and</u> maintain relatively stable indoor air temperatures. Through careful design, this can be accomplished without the use of excessive amounts of thermal mass material and glazing.

BIOGRAPHICAL SKETCH

Ed Mazria is an architect and professor at the Department of Architecture, University of Oregon in Eugene, Oregon. Mr. Mazria has developed computer programs for passive solar designs.

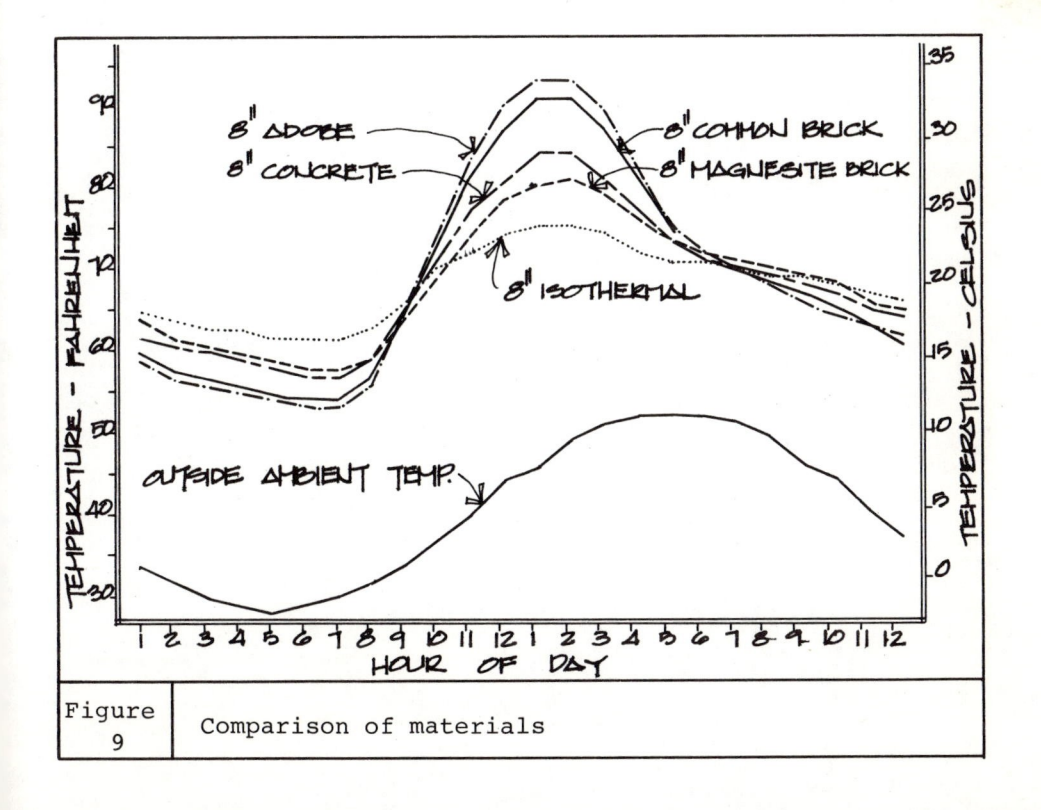

Figure 9. Comparison of materials

ENERGY-PROCESSING BUILDING MATERIALS

Presented and prepared by: Day Chahroudi

This paper discusses three novel energy-efficient building materials and two building designs that exemplify their uses.

MATERIALS
We have develped a vocabulary of three complementary building materials specifically for passive systems. These materials are:
    1. Transparent Insulation* materials for use in reducing heat loss in windows, skylights, greenhouses and solar collectors. The primary components in these materials are "heat mirror" coatings which are transparent to solar radiation and reflective to long-wave infrared (heat) radiation.
    2. Optical Shutter* materials which vary their transmission of solar radiation as a function of temperature. Such materials can be used in windows, greenhouses and other glazing to prevent overheating from solar heat gain.
    3. Thermocrete* structural concrete materials filled with a phase-change material for heat storage. This material acts thermostatically and is ideal for distributed heat storage and passive solar heating.

These three materials allow the utilization of local natural
---
* patented Suntek products

energy resources for space conditioning buildings in a manner that is completely integrated with their architecture. Since these materials are designed on a "molecular" level, they have no moving parts. Since they replace the ordinary components of a building, their energy functions are purchased for a marginal increase in cost.

The heat mirror, which is coated onto both sides of a plastic film, has a transmission of solar energy of 81% and an emissivity of 0.11. When used in conjunction with dead air spaces on both of its sides, its thermal conductance is 0.13 Btu/ft^2 F. It is the highest performance Transparent Insulation (excluding expensive evacuated systems) yet developed.

The Optical Shutter consists of a thin layer of a thermo-optical material that is laminated between two sheets of glass or two films of plastic. This material has the property of turning from completely transparent to an opaque white with a transmission of solar energy of 15% when heated above its transition temperature. This change of stae is thoroughly reversible and occurs over the narrow temperature range of 3°C. Due to this thermoactive property, the optical shutter is able to regulate diffuse energy transactions with local feedback. Thus, it makes it possible to build structures that are homeostatic around the human comfort zone.

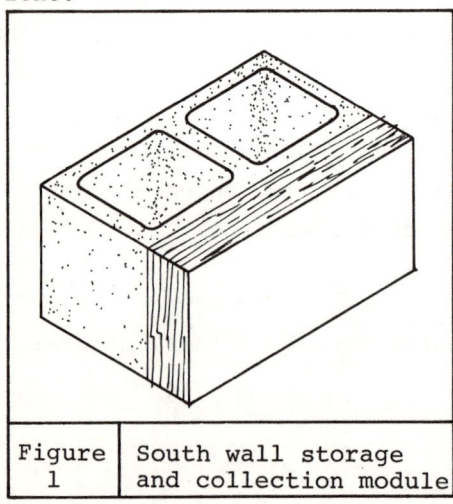

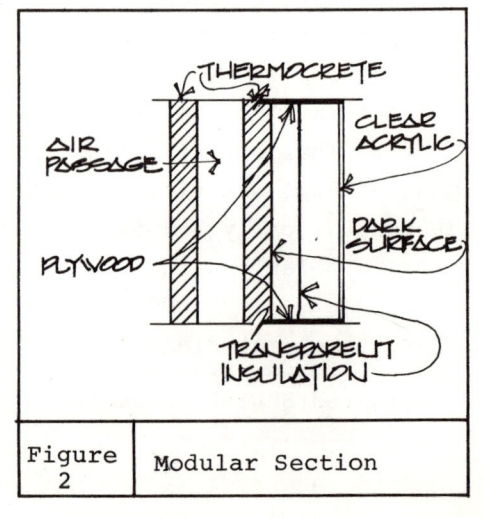

| Figure 1 | South wall storage and collection module |
| Figure 2 | Modular Section |

Solar Architecture

Thermocrete, another material developed by Suntek, consists of a phase-change thermal storage material permeating a porous cement matrix which is sealed with a plastic coating. This material is packaged as an inch-thick tile for distributed storage in walls or ceilings and as a concrete block with the same strength as conventional concrete blocks for structural uses. As this material stores heat at its melting point, it can also handle energy fluxes in a distributed homeostatic fashion utilizing local feedback.

SOLAR WALL

The following building design is a refinement of an MIT solar house built in 1954 and the later Trombe wall. Precast blocks of Thermocrete can be used as the structural and thermal storage elements in a solar wall constructed from modules that integrate wall structure, solar energy collection, thermal storage, thermal regulation and thermal distribution. Combining these five functions in one modular element greatly simplifies solar heating and thereby drastically reduces costs. We believe that the system presented here is the most economical solar heating scheme yet proposed. Since it has no moving parts, reliability is exceptionally high while maintenance is essentially zero.

CLIMATIC ENVELOPE

A climatic envelope is a radically different approach to the utilization of solar energy for space conditioning buildings. A complete integration of the structural and energy functions of a building results from using energy-handling building materials. The climatic envelope separates the weather protection functions of a building from the use patterns. Wind, precipitation, sunshine and the extremes of temperature

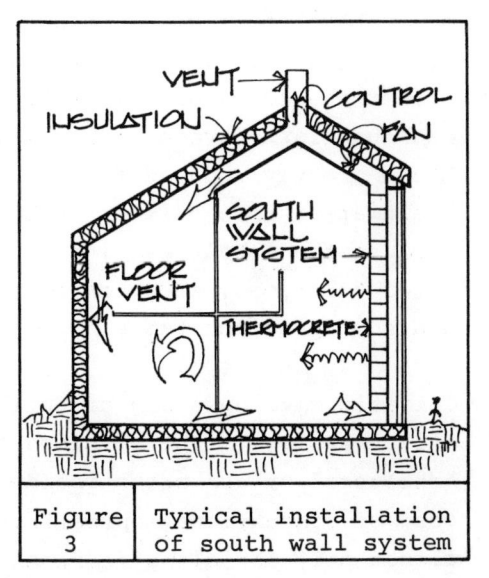

| Figure 3 | Typical installation of south wall system |

are mediated by the Transparent Insulation and Optical Shutter which acts with variable selectivity to maintain good weather in its interior. The internal architecture is relieved of structural demands and can be as sculptural or as unassuming as the situation calls for. The simplest example is an inflated bubble fabricated from thin transparent plastic films. A dome, say, that comes in a box, is inflated, heat storage and/or backup heating and cooling systems are installed, and the weather inside is always good. A dome made out of the solar membrane would supply over 85% of its own heat in the southern 3/4 of the United States with a three-day storage system.

It is felt that by extending the good weather area of a building beyond its living and use areas, by providing a pleasant transition to the outside environment, and by eliminating the bad weather areas between buildings, definite relaxations will take place in the occupants. Removing the weather and structural restraints from the living and use areas inside buildings should produce a corresponding relaxation and personalization of the architecture, not to mention a great reduction of first and running costs.

For a suburban development the functions of a home under a climatic envelope include mainly visual and audio privacy and pleasing apprearance and storage space. The envelope can be either one for each family, separated by berms or the envelope can span large areas with families separated by trees and fences. These greatly reduced functions are much more within the realm of the amateur home builder in both skill and financial resources. Because the "shelters" inside the climatic envelope have such reduced functions, they can consist largely of movable partitions and soundproof curtains.

Climatic envelopes can be built entirely with existing materials and technology. The combined first cost of the envelope and the "shelters" or use areas inside it will be lower than the first cost of conventional operating costs. Activities at present enclosed only for climatic reasons could take place in the "open air." An inter-connected assembly of variously shaped and sized envelopes with both internal and external spaces and structures could completely revolutionize urban design and have a large impact on life styles and social forms.

BIOGRAPHICAL SKETCH
Day Chahroudi is a designer with Suntek in Corte Madera, California. Mr. Chahroudi is also a member of the Board of Advisors of the Roaring Fork Resource Center, Aspen, Colorado.

Solar Architecture

The different materials used were gold colored aluminized mylar, polyslim mylar needle punch, and a vacuum deposited foil on a nylon scrim (commonly called foylon). In the first test, a curtain with the following profile: gold mylar (shiny side in), 3/4" air space, needle punch, 3/4" air space, gold mylar (shiny side in), 3/4" air space, needle punch, and gold mylar (shiny side out) was raced against a 1" piece of foam with an R-value of 4.1. After six hours, the curtain's water was hotter, containing 150 more Btu's.

In the second test, the needle punch was replaced with foylon and the curtain raced the same piece of foam. After seven hours, the curtain's water was hotter, 350 more Btu's. In the third, fourth, fifth and sixth tests, the middle sheet of mylar was replaced with foylon. In the third test, the curtain was raced against the same 1" piece of foam. After seven hours, the curtain's water was hotter, containing 500 more Btu's. In the fourth test, the same curtain was raced against a 1 1/2" piece of foam with R-value of 6.15. After nine hours the curtain's water was hotter, containing 350 more Btu's. In the fifth test, the same curtain was raced against a 2 1/2" piece of foam with an R-value of 10.25. After six hours, the curtain's water was hotter by 200 Btu's. In the sixth test the same self-inflating curtain raced a 4" piece of Johns-Manville fiberglass batt insulation with an R-value of 14. After six hours, the curtain's water was hotter, containing 100 more Btu's.

Experiment	Time (hrs.)	Curtain vs. Control	Margin (Btu)
1	6	5 layers vs. 1" foam	150
2	7	5 layers vs. 1" foam	350
3	7	5 layers vs. 1" foam	500
4	9	5 layers vs. 1 1/2" foam	350
5	6	5 layers vs. 2 1/2" foam	200
6	6	5 layers vs. 4" fiberglass	100
Table 1	Curtain Test Data		

METHOD OF CALCULATIONS

Hq = heat loss, Btu/hr/ft^2/F
U = heat transfer coefficient, Btu/hr/ft^2/F
A = surface area, ft^2
ΔT = change in temperature

Hq/Btu (5 gallon can) = (Thermal Mass)(ΔT)

To find insulation values:
Hq (Foam box) = (U) (Area of Box) (ΔT air)
Hq water - Hq foam box = Hq unknown material
Hq unknown material = (U) (Area of unknown material) (ΔT air)

$$U = \frac{Hq \text{ unknown material}}{(\text{area unknown material})(\Delta T \text{ air})}$$

CONCLUSIONS

In the tests, the comparative values were stressed. Each curtain tested was a sandwich of four 3/4" air-spaces and five layers of material. The foylon proved to be a better insulator than the needle punch. In all of the tests, the curtain proved better than the control. In the last test, the curtain proved to be a small amount better than 4" of fiberglass insulation with an R value of 14, which agrees with the value supplied by the ASHRAE Handbook of Fundamentals. The tests prove and confirm that the curtain will be a successful movable insulation.

BIOGRAPHICAL SKETCH

Ron Shore is a solar consultant and teacher at the Colorado Rocky Mountain School (Solar Applications Laboratory) in Carbondale, Colorado. Ron is also a member of the Board of Advisors of the Roaring Fork Resource Center. James Gronen is a student of Ron's.

NORTHERN WINDOWS AND SOLAR ARCHITECTURE FALLACIES

Presented and prepared by: Raymond N. Auger

This paper presents the results of a simple experiment that contradicts some of the current ideas of solar architecture. These experiments utilized a small meter to measure solar intensity on a south-facing hillside adjacent to Aspen, Colorado. On a cloudy day, within two to three hours of solar noon, the intensity of solar radiation in Aspen is around 50 Btuh/square foot and is fairly uniform throughout the sky. At noon it may be 75-85 Btuh/square foot facing south, and 55-60 looking north, east or west. On a fully sunny day it may be about 365 Btuh/square foot at an angle normal to the sun. In the winter it may go as high as 385 Btuh/square foot because of reflection from the snow. Consequently, cloudy-day radiation intensity is 1/7 of the direct sun radiation.

On a cloudy day in May 1977, at 10 am, the northern radiation was approximately 55 Btuh/square foot at a tilt angle of 45 degrees. On a clear sunny day, the northern radiation dropped to 15 Btuh/square foot. The apparent reason is that on a sunny day, some reflected radiation comes from the mountain side which is covered with scrub oak and exposed earth. It is not a highly reflective surface, but obviously there is some ground radiation. The meter is in full shade on the north side of a house.

CALCULATIONS

The calculations for the experiment use a double layer of plastic glazing (ribbed, extruded acrylic) with an advertised U-value of 0.62. This material only has an 80-85% transmittance as claimed by the manufacturer. By using two layers of this material (four layers of glazing), the overall transmittance is 65-70% and each air space would have a U-value of 0.62. Assuming an overcast sky radiation of 50 Btuh/square foot above an insulated black surface, what is the stagnation or equilibrium temperature? Surprisingly, the calculations showed a temperature of 90 F above ambient air temperature. Assuming that a residence has a north-facing window with four layers of plastic glazing with insulated black walls that limit heat loss to that of an R-8 surface equal to the window size, the interior temperature would be 90 F on an overcast day with an ambient air temperature of 0 F.

TEST MODEL

To verify these calculations, a small test model (Figure 1) was built. The model consisted of four-domed layers of Lexan over a sheet of aluminum backed with a sheet of two-inch styrofoam. A thermometer was attached to the aluminum and the model faced north at noon on a day when the air temperature was about 30 F. After one-half hour the thermometer read 90 F. The Lexan was very dirty, so the transmittance through each layer was probably down 10-15%. Despite the 10-15% loss through each of the four layers, the surface temperature of 90 F confirmed the calculations and stimulated a

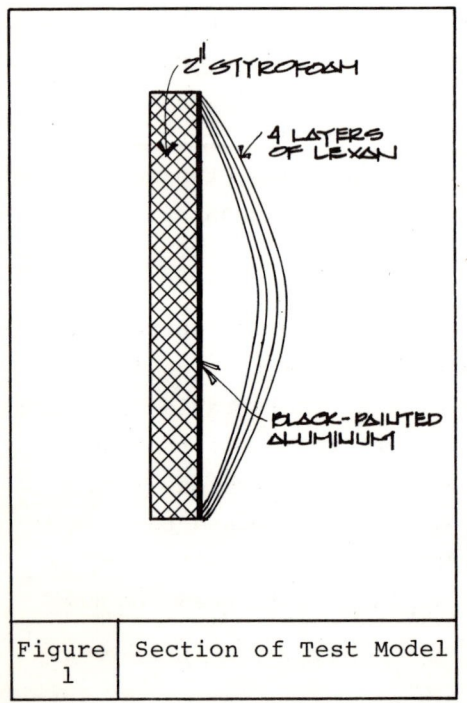

Figure 1 — Section of Test Model

great deal of speculation. Suppose instead of having dirty Lexan, a clear Teflon which only has a 5% transmission loss is used with a better insulation (R-16, R-20 or R-30) behind the absorbing surface. What is the equilibrium temperature? Clearly, it will be well above 100 F over the ambient air temperature.

The implication of this experiment is that a solar house should have its south wall used primarily for solar collection, putting heat into storage, and the north, east and west walls should be glazed with many layers (at least four to six). This would indicate that from about 10 am to 5 pm on the cloudiest of days there would be more than enough heat inside the house from the radiation through the windows. Ventilation would be necessary on almost any day of the year. At night, losses would be kept down by the number of layers of glazing and perhaps by movable insulation. As many as six or seven layers of glazing might be desirable to keep night losses to a minimum.

Most solar heating systems have large storage capability to provide heat during cloudy days. However, a house designed with multi-layers of north-facing glazing will have heat on cloudy days, minimizing the need for heat from storage. On overcast days, the only period in which you would have to extract heat from storage would be from 5 to 6 pm through bedtime. On clear sunny days, some direct radiation through the south wall could offset the drop of heat input through the north wall.

CONCLUSION

The overall conclusion is that the concept of passively solar heating a structure with north, east or west windows is practical, and that the commonplace notion of eliminating northern windows (or east and west windows) is erroneous and is based on the assumption that two layers of glazing is maximum. An economic analysis of the northern-heating concept would put the cost of storage and collectors against the cost of multi-layers of glazing and possibly nighttime movable insulation. The aesthetics of northern light must tilt such analysis towards the use of these windows. Although on a Btu/dollar basis, the collector/storage system probably comes out ahead for climates with minimum cloudy days.

Perhaps the most important conclusion is that the 100% solar is a practical objective, regardless of the duration periods of overcast. With enough of the right kind of glazing, flat-plate panels can be built which will heat domestic water to over 100 F on heavily overcast and cold days. There is no question that it is more expensive to collect heat when the radiation level is 1/7 that of direct sunlight. However, it is less expensive than attempting to collect and store enough heat from direct sunlight to compensate for seven days without any solar heat input. The optimum solar home design must balance aesthetic factors with the utilization of direct and diffuse radiation as normally experienced at the site, with multi-layers of glazing as the key element in diffuse radiation collection.

BIOGRAPHICAL SKETCH

Raymond Auger is an author, inventor and energy economist in Aspen, Colorado. He has been actively developing innovative solar hardware and systems. Mr. Auger is also a member of the Board of Advisors of the Roaring Fork Resource Center in Aspen.

HEATING REQUIREMENTS FOR BUILDINGS

Presented by: William R. Harmon
Prepared by: Gregory Franta

The procedure for calculating the heating requirements for buildings may be quite complex and it is often difficult to accurately determine the numerous variables. This paper is a summary of the steps often used to determine the heating requirements of buildings. Further description and definition may be found in mechanical and electrical systems textbooks[1] or the ASHRAE Handbook of Fundamentals.[2]

STEP 1. Collect local weather data. This is often available from a nearby weather station, local engineer, utility company or the data may be interpolated from the Climatic Atlas of the United States.[3] Average and extreme monthly temperatures and wind velocities are the most important data to collect for heat loss calculations.

STEP 2. Determine the area of the exterior surfaces for each room. Break these into the areas of different composition that would indicate a different thermal resistance (i.e. walls, windows, roof, etc.).

STEP 3. Calculate the U value for each of the areas determined in Step 2. The U value is the overall coefficient of heat transmission (usually expressed in Btuh/sq ft F). The U value is the reciprocal of the total thermal resistance (R-value). R-values for various materials may be determined from charts in the ASHRAE Handbook of Fundamentals or given by the manufacturers of various building products. Figure 1 illustrates a typical procedure used for determining the U value of a sample wall section.

STEP 4. Determine the design temperatures. These outside design temperatures are determined in Step 1. The inside design temperatures are generally 68 F. However, lower temperatures are often assumed in less occupied rooms.

STEP 5. Calculate the conduction heat loss for each room. This is done by using the following equation: $Q_c = A \times U \times \Delta t$, where:

Q_c = conduction heat loss (Btuh)
A = Area of surface (square feet)
U = Coefficient of heat transmission (Btuh/sq ft F)
Δt = Difference in temperature from inside air to outside air ($\Delta t = t_{outside} - t_{inside}$) (F)

The conduction heat losses for each exterior surface must be totalled.

STEP 6. Calculate the infiltration loss for each room. Typically the calculated infiltration losses in walls are losses through the cracks in doors and windows. This loss is usually calculated by the following expression:

$Q_I = AC \times 0.018 \times \Delta t$

NO.	ITEM	R
1	EXTERIOR AIR FILM	0.17
2	WOOD SIDING W/ BACKBOARD	1.40
3	6" FIBERGLASS INSULATION	19.00
4	½" GYPSUM BOARD	0.45
5	INTERIOR AIR FILM	0.68

TOTAL R = 21.70
U = 1/R = 1/21.70 = 0.046 Btuh/ft²/F

Figure 1	Typical procedure for determining U values

Solar Architecture 117

where:

Q_I = Heat loss through infiltration (Btuh)
AC = Number of air changes per hour (cu ft/hr)
0.018 = The product of the average density and specific heat of air
Δt = Difference in temperature from inside air to outside air ($\Delta t = t_{outside} - t_{inside}$) (F)

The following are assumed common air changes:[4]

Room Type	Air changes/hr/cu ft
No windows or exterior doors	1/2
Windows or exterior doors on 1 side	1
Windows or exterior doors on 2 sides	1 1/2
Windows or exterior doors on 3 sides	2
Entrance halls	2

STEP 7. Calculate the total heat loss. This is done room by room using a format similar to that in Figure 2. After the heat loss for each room has been calculated, the total heat loss for the building is determined by adding the heat loss for each room together.

The total heat loss represents the heating requirement (Btuh) for buildings under extreme conditions. Therefore, the boiler or heat source must be able to provide this amount of heat to keep the building at 68 F under the specific outside design

SPACE	ITEM	AREA OR AIR CHANGE	U-VALUE OR OTHER	Δt (F)	Q (Btuh)
BEDROOM	WALL #1	40	.046	85	156.4
	WALL #2	70	.052	85	309.4
	ROOF	200	.050	85	850.0
	WINDOW	15	.560	85	714.0
	INFILTRATION	1980	.018	85	3029.4
			TOTAL HEAT LOSS →		5059.2

Figure 2	Typical format for calculating heat loss on a room-by-room basis

conditions. This procedure is only a summary and further details can be found in the references listed earlier. Also, for a more accurate assessment of the heating requirement, the solar heat gain and storage needs to be calculated using daily design conditions. The ASHARE GRP 170[5] is a good reference for these calculations. Other variables will affect the heating requirements, and it should be noted that this procedure only estimates the heat loss for buildings.

BIOGRAPHICAL SKETCH

William R. Harmon is a consulting engineer with Beckett-Harmon Associates in Denver, Colorado. Mr. Harmon is also a member of the Board of Advisors of the Roaring Fork Resource Center, Aspen, Colorado.

ENDNOTES

1. McGuinness, William J. and Stein, Benjamin, Mechanical and Electrical Equipment for Buildings, John Wiley and Sons, New York, NY, 1971.
2. ASHRAE HANDBOOK OF FUNDAMENTALS, American Society of Heating, Refrigerating and Air-Conditioning Engineers, New York, NY, 1976.
3. Climatic Atlas of the United States, US Department of Commerce, National Climatic Center, Asheville, NC, 1974.
4. McGuinness, Ibid., p. 127.
5. ASHRAE GRP 170 Application of Solar Energy for Heating and Cooling of Buildings, American Society of Heating, Refrigerating and Air-Conditioning Engineers, New York, NY, 1977.

GREENHOUSE CONSTRUCTION WORKSHOPS "BARN RAISING STYLE"

Presented and prepared by: William F. Yanda

In the last year, the Solar Sustenance Project has coordinated eighteen workshops where the primary goal was to build a solar greenhouse with residents of local communities. The workshop format provides a unique opportunity for education by demonstration at a variety of levels. In New Mexico the project was funded by the State Energy Resources Board and the emphasis was directed toward the low/fixed income sector, although people of all incomes and educational levels participated.

The dynamics of each workshop were as unique as the local communities themselves. However, in general, the format was as follows: (1) Interested community organizations (community action programs, local solar energy groups, social service organizations, etc.) were contacted by the director and informed of the project and its goals; (2) The workshop dates, usually a weekend, were established; (3) A package containing organizational plans, a materials list, site criteria and media contact information was sent to the local coordinators. A great deal of the responsibility for the success of the workshop was put into local hands. By doing this, local involvement was stimulated and many of the problems of "outside" inspired programs were avoided; (4) At a public meeting the principles, examples and options of working greenhouses were explained in an hour-and-a-half lecture/slide show presentation. Each public meeting was attended by an average of seventy-five people. Besides general education, an important function of the meetings was to inspire the audience to come out and build the greenhouse the following two days. An unplanned benefit of the public meetings was that in several of the locations (Gallup, Farmington, Portales) the nucleus of a local solar energy association and information exchange forum was established; and, (5) The greenhouse was built by participants from the community. It was during this time that the many levels of education were taking place at once. Often, people who had studied solar applications for several years had never actually placed hammer to nail. For them, the building aspect was particularily important because the realities of construction often take precedence over theoretical simulations (i.e., if the optimum intersection angle of two planes is $47\frac{1}{2}°$, most carpenters will interpret it to 45°). For others familiar with building, the important questions related to materials, costing and specific site problems. From this group came crew chiefs who would act as foreman and educators for those less acquainted with construction. Anyone who wanted to work was given a job. During the two-day building session, many interested non-builders dropped by the site to ask questions and check on the progress of the workshop. It is estimated that an average of ninety people attended each activity. For the twelve New Mexico workshops, that would mean 1,080 people directly involved in the program. The end of the two-day construction saw the demonstration 75-98% completed. The owners of the greenhouses and interested friends completed finishing work, interior layout and planting.

It was of particular interest to the project to determine if a "multiplier effect" in the private sector could be obtained by the government sponsored demonstration solar units placed directly in the communities. To obtain data, 100 cards were sent to workshop participants chosen at random. The following choices were presented: (1) I have built an attached greenhouse; (2) I plan to build one this year; (3) I plan to build one. At this writing, one week after the cards were sent, 33 have been returned. The responses are as follows: (1) 8-24%;

Solar Architecture

(2) 12-36%; (3) 10-30%; and, (4) 3-9%. Even assuming that some of the people who have built were planning to build anyway, and that many of those who plan to build, won't, this is still a robust result. It indicates that eighty solar greenhouses (ten times the sampling survey) have already been built and that many more will be built within the year as a result of the New Mexico project.

Many people, particularily elderly people, cannot build an attached greenhouse by themselves. It was a natural outcome of the workshop format that in several communities (Las Vegas, Alamagordo, Taos) participants who wanted to build for jobs were put in contact with customers who wanted greenhouses built. These small job builders learned greenhouse building techniques at the workshop and have contracted jobs in their own communities through the project.

Of course, there are drawbacks to the "short-time frame" workshop program. For instance, it isn't possible to discuss every possible option in building and operation when the prime objective is to finish the majority of construction on the demonstration unit in two days. Some people with sincere questions are sure to be missed. Another difficulty is in having so many people show up that they are tripping over one another. Safety and quality control can become strained in this situation.

Because of the advantages of direct contact with the people and the increasing educational need for owner built systems, the construction workshop is a viable medium for many solar applications.

BIOGRAPHICAL SKETCH
William F. Yanda is the director of the Solar Sustenance Project in Santa Fe, New Mexico. He is very active in the design and construction of attached solar greenhouses. Mr. Yanda is also a board member of the Roaring Fork Resource Center in Aspen, Colorado.

ENERGY FLOWS IN THE GREENHOUSE

Presented and prepared by: Herbert A. Wade

Anyone who has been in a greenhouse of any type on a sunny, cold winter's day knows first hand the power of the sun. More often than not, powered ventilation systems are needed to rid the structure of the great excess of energy pouring in. Unfortunately, if a visit to the same greenhouse were to be made on the following cold, clear night, it is probable that large energy losses would be evidenced by the necessity for a large amount of heat from a unit fired by electricity or a costly fossil fuel. What a discouragement in these times of costly and increasingly scarce energy resources to have to use power to get rid of excess heat in the daytime and power to add heat at night. If only it were possible to store up the excess in the day and release it in the night. Of course, it is possible to do just that granting a little insight into the functions of a greenhouse and the energy flows that occur in the structure.

To begin, one must understand the energy needs of a greenhouse. Why is the greenhouse there in the first place? Usually, a greenhouse is designed to provide the proper environment for plant growth and production. Thus the energy flows are a result of the needs for those plants. Of course, each plant type requires a little different environment for best production, but basically the needs are ones of moderate temperature, water access, carbon dioxide, oxygen, mineral nutrients and light. The two energy-related needs are, of course, the temperature and light requirements. The temperature must be high enough to allow growth but not so high as to cause metabolic change. The light, being the activating energy for growth, must be present in amounts generally too high to be inexpensively provided by electrical illumination. With these two factors in our minds, let us examine the "conventional" greenhouse design which is essentially a space enclosed completely by glass or plastic. During the day, the sun with its direct solar energy and the sky with its diffuse energy component provide more than enough light within the greenhouse for growth except on very cloudy days when plant metabolism may be limited by the relatively low-light levels. When the plants convert this energy to their uses, over 90% is converted into heat thereby warming the plant and the air around it. Of course, the air absorbs but little heat without rising significantly in temperature and then physically rising to the higher levels of the building. In a very short time, the greenhouse has accumulated a layer of hot air at the top and rapidly becomes too warm at the level of the plants as well. Ventilation is needed. Likewise, at night the poor insulative character of the transparent skin allows the air to cool rapidly and excessively requiring heaters to keep the plants warm enough to grow properly. The interesting thing, from an energy standpoint, is that the excess of energy in the daytime is almost enough to get the greenhouse through the night if it could be trapped and released at will. Secondly, the amount of light available on all but overcast days is reduced only a few percent, if the north surfaces of the building are shaded; particularly if the greenhouse is laid out with its long axis east-west. From this it is not a difficult step to improve the conventional greenhouse performance.

The first step is to make the north wall and much of the roof solid and well insulated. The energy entry is reduced but little but the losses are reduced by nearly half, in fact, if

the interior of the solid wall is light in color the energy at
the plants may go _up_ because light that formerly either passed
all the way through the greenhouse from south to north or was
reflected out through the north wall and roof now is trapped
inside and reflected back onto the plants. Naturally with losses reduced, the temperature goes up and more ventilation is
needed to keep things moderate for the plants. Obviously, the
insulated walls allow lower losses at night so the heater does
not have to work quite as hard. However, the energy balance of
the greenhouse has not improved much yet, although the cost of
construction may have increased significantly. The problem of
having an excess of available energy in the daytime and too
little at night is still present. Therefore, the heat should
be stored.

Why does it get so hot in a greenhouse so fast? The main reason is one of heat capacity. When the incoming solar radiation
strikes plant surfaces there is little mass there and after
removing a few percent of the energy for metabolic needs, the
leaf simply gets hot and passes its energy off to the air
around it. Of course, air also has little capacity to hold
heat and rapidly increases its temperature resulting in a
"hothouse." If the sunlight strikes a surface which has a large
heat capacity and energy is absorbed, the temperature rises much
more slowly. Naturally, the sunlight is striking all the surfaces in the greenhouse either directly or after one or more
reflections. If these surfaces are high in heat capacity, then
they will effectively store heat. This storage capacity reduces
the energy that is passed to the surrounding air by not getting
hot immediately. When the sun goes down, the plants and air
will cool rapidly, and the heat stored in high-heat capacity
materials will cool slowly, helping to maintain energy levels
above the outside ambient. We now have a greenhouse well on its
way to becoming energy independent. The following is a summary
of what has been done:

    1. Reduced glazed area to only that needed for adequate
       light entry.

2. Provided high-heat capacity materials to absorb and store excess energy in the daytime and release it at night.
3. Thoroughly insulated the non-glazed building surfaces.

Interestingly enough, the resulting design is the basis for most successful greenhouses around the world prior to the availability of low cost fuels. Examples of designs using the above principles were present in the 17th and 18th centuries in the Netherlands and other parts of the Continent. The completely glazed greenhouse is more recent than that and probably evolved in areas which had few clear days in winter and needed all the light they could get and were close to supplies of coal. The fact that such greenhouses are also found in such unlikely places as Phoenix, Arizona and Santa Fe, New Mexico, is more a tribute to cultural inertia than intelligent design.

By carefully analyzing the needs of our plants, the energy environment of the site and our use patterns we can "fine tune" the three basic design components and produce a greenhouse optimized for our localized, individual need. Let us examine each item separately.

To glaze or not to glaze, that is the question. The main reason for glazing at all is to let light in for the plants. Theoretically, we can do better at temperature control, if we build the structure with no glazing at all, collect heating energy some other way and duct it into the structure. We really cannot afford to do that in most places, since light is more expensive than heat. The way to look at it is to start with some small window, say 2' by 2' and increase its size mentally and for each increment relate the heat loss potential to the light gain potential in a specific climate. This results in a lot of additional light for a given amount of heat loss, but as the glazed area gets larger, the gain begins to be largely offset by the loss. When the east walls, west walls and the roof are glazed, climates with a lot of clear days will show that the added glazing does little for light entry but greatly increases heat loss and is not justified. For climates

Solar Architecture

the day without limited light entry or excessive heat loss.

Which storage is better; rocks or water? Well, before examining that somewhat emotionally charged question, perhaps we should think about how the storage system does its job. We have established that we want tight coupling between the energy storage system and the energy source. Now if the sun is shining directly on the storage device, energy flow rates are much higher than if it is not. Thus the storage which is in the sun must be able to get the heat away from its surface rapidly, otherwise the surface will get much hotter than the interior, and the coupling is poor. On the other hand, if the storage is exposed primarily to hot air as the energy source, flow rates are much lower and poorly conducting masses can be as good as rapid ones at getting the heat out of the air and into the storage.

Now we can ask: water or rock? Water contained in thin-walled containers can accept the high energy flows of direct sunlight and rapidly distribute that energy through its mass by conduction and convection while rock must rely on conduction alone. So for storage which is directly illuminated by the sun, barrels, cans, jugs, troughs of water will couple closely to the energy source while rock will not do nearly as well. On the other hand, for the absorption of energy from the air inside the greenhouse which has been heated by the plants and the non-storage surfaces, rock is nearly as good as water containers (since the rate of transfer is of the same order as the energy applied). Additionally, loose rock can have air circulate through it and can have a much larger surface area exposed to air flow than water in economically sized containers. The rock can couple tighter to energy contained in air than most water storage systems. So, if our design places water storage in the direct illumination area and rock or masonry elsewhere, we may have the most economical, thermally satisfactory system.

An area of potential for heat storage often ignored is that at the peak of the greenhouse. There the air is the hottest, losses are the highest and the need for ventilation is strongest. It is clear that storage placed so that rising hot air will have to pass by (or through) it will do a better job of extracting energy than one placed vertically along the wall. If economically feasible, jugs of water hung from the peak, ceilings of poured concrete with exterior insulation, overhead racks of rock or other methods of placing heat storage <u>high</u> in the structure should be considered. For the non-passive systems purist, a small fan drawing this heated air down from the peak and blowing it around the storage system (or through, if possible) will achieve the same ends without the intimidating influence of a few tons of rock overhead, as you walk through tending the asparagus. However, the overhead storage has other advantages as well, this brings us to the other side of the coin: getting the heat out at night.

How does the storage system work during the release of its stored heat? Infrared radiation seems to be the answer; at least for a greenhouse that has little infiltration problems. Energy is also released through conduction and convection. But since all parts of the greenhouse interior except the glazing have the characteristic of very slow heat flow to the outside (e.g. well-insulated and/or highly absorbing in the long wave infrared such as plant surfaces, painted surfaces, earth, etc.), then energy movement through instantaneous radiation occurs and comes to a balance long before conduction and convection have time to transfer much heat. The exception is the glazing. Convection cells are formed under the glazing and do cause a layer of cold air to form near the ground which is warmed by contact with that ground and other energy-storing masses nearby. Radiation is still the most significant aspect of the interior energy flow for the plant, provided it is a foot or more above the floor. This is evidenced by the fact that passive greenhouses show very little temperature differences from high to low until very near the floor. If convection and conduction from storage were the major means of moving heat from storage, then there would be a significant thermal stratification from top to bottom. In a greenhouse of conventional design with little

Solar Architecture 131

storage, there is major stratification at night when the heat is needed.

The fact that long-wave infrared radiation is the major method for heat to be released from storage may affect our geometrical designs. The plant, for best health, should see equal energy flows from all sides; thus storage on the east, west, north, overhead, below and even the south may be preferred. One way of doing this without covering all the interior surfaces with beer cans full of water is to use infrared reflecting surfaces where there is no storage. A surface which absorbs the infrared and re-emits it is equivalent to a reflector. So an insulated surface of almost any material will work, as long as its temperature can come to that of storage without a lot of heat flowing to the outside. The only surface which we cannot make do that is the glazing itself. It will absorb the radiation and rapidly conduct it through to the outside. Also, it will stay much colder than that of storage. If we can arrange for an infrared reflecting surface (or a well-insulated surface) to cover the glazing, then we will give the plants equal energy flows on all sides and make optimum use of our storage. Naturally, a well-insulated cover is better than a sheet of aluminum foil for covering the glazing at night; but the sheet of aluminum foil is better than nothing at all, and for many climates will be entirely adequate. It should be noted that for some climates and some plants, no night treatment of the glazing is needed, but that is generally not the case.

One final aspect to be considered about energy flows in greenhouses relates to the needs of the plants. For best absorption of light energy in the storage system, we should use black in color. The problem then arises that the plants will "see" a markedly different illumination environment on the side where storage occurs related to the glazed side. This may cause growth problems and may cause plants near storage to grow differently from those near the glazing. On the other hand, if we paint our storage white, the plants love it, but direct absorption of sunlight by storage is minimal. One thing we can do is take advantage of the fact that the plants do not need equal amounts of all the spectrum of colors. The plants we commonly find in domestic or commercial greenhouses need red and blue light for growth. Near infrared and green are poorly utilized, if at all. Thus if we paint that portion of our storage system which the plants see either red or blue, the storage will absorb the rest of the radiation from the sun and store it. The poorest color next to black would be leaf green.

It has been demonstrated over and over that the passive solar greenhouse works and can be a highly productive, low-energy cost system. Unfortunately, it has been demonstrated also that they can be designed to fail. The key to success is understanding the energy flows and putting them to work for you and your plants in your climate. No one design will fit even a fraction of the possible universe of climates and plant needs. If we are willing to add some simple active elements to our design, we broaden the design applicability. Nonetheless, we need to understand our climate, the greenhouse energy flows, and yes, our own needs and motives before we can "harvest the sun."

BIOGRAPHICAL SKETCH
Herbert A. Wade is the Deputy Director of the Arizona Solar Energy Research Commission in Phoenix, Arizona. Mr. Wade has been actively involved with the design and construction of solar residences in Arizona.

WET-DIRT STORAGE FOR
A SOLAR GREENHOUSE

Presented and prepared by: Joseph B. Orr

BACKGROUND

During 1975, in collaboration with Mr. William B. Edmondson, publisher of the periodical, Solar Energy Digest, the Orr family constructed in Laramie, Wyoming, at an elevation of 7,200 feet, an experimental pilot or prototypal house-attached solar-heated greenhouse, using an innovative and cheap means of storing heat in a wet-earth heat bin under the greenhouse garden. An early version of our design was described this same year in the March issue[1] and a modified version in the August issue[2] of that publication. Prior to construction the only published material on wet-earth heat storage known to us were two inspirational designs neither of which had been reduced to practice: (1) one page in Domebook Two[3], published in 1971, presenting Day Chahroudi's "biosphere" as an integration of greenhouse and living quarters under one roof with a wet-soil solar heat battery under the garden; and (2) in the November 1973 Solar Energy Digest[4] with Edmondson's fabulous "Solterra Home" with its 58-days' worth of wet-earth heat storage under the house's main floor.

We cannot boast that wet-dirt storage has been endorsed by a wave of popular acclaim. To our dismay, Chahroudi's later biosphere, illustrated in the July 1974 issue of Equilibrium[5], abandoned his former wet-soil battery for a massive storage wall.

We wondered with some misgivings why he had forsaken wet dirt. Indeed, hardly anyone was backing our cause. The feasible merits of wet dirt is omitted from the writings of venerable pioneers in the field. Farrington Daniels' chapter on heat storage in Direct Use of the Sun's Energy offers nearly everything under the sun as possible heat storage; from water and rocks to salts, from metals, mines and caves to water ponds with black bottoms--but neglects to mention wet dirt. Maria Telkes in her chapter entitled "Solar Heat Storage," appearing in Solar Energy Research,[7] discusses rocks, water and salt hydrates--but nothing on wet soil. None of the dozens of new books, which we have examined, says a word on the subject. Of all 319 buildings scattered over the world, described in William A. Shurcliff's 13th edition of Solar Heated Buildings, A Brief Survey,[8] not one (unless we missed with our count) uses wet-dirt storage in an active system. Bill Yanda in his new book[9] depicts a house designed by Lee Porter Butler where the storage partially consists of 18 inches of dirt under the floor. A paper from Cornell University presented at the "Solar Energy - Fuel and Food" Workshop[10] in April 1976 at Tucson briefly discusses an earth-air heat exchanger located beneath one of their experimental greenhouses; the feasibility of which has yet to be determined. Lastly, we have been in communication with Frank D. Werner of Jackson, Wyoming who, with Richard C. Greig, have devised a promising method of heating up earth slabs from intervening trenches of rock serving as exchangers.

And this--to the best of our knowledge--is the meager state of the art and literature of wet-earth heat storage. Among our informants, William Edmondson was the sole wet-dirt advocate. We worried about this. Was everybody else amiss? And were we so right? Wet dirt is so irresistably simple. It is so mathematically reassuring. It is so abundantly and readily available. It is so undeniably economical. Then why are not others using it? The question is yet unanswered.

WHY WE BUILT IT
At Laramie, where we live high on the steppes of Wyoming in

Solar Architecture   135

America's own Outer Mongolia, we have a very short growing season. Though once optimistically rated longer than sixty days, with the erratic and freakish weather we have had in late years, we have come to accept the lesser expectation of a reliable frost-free gardening season of (with luck) about 10 to 20 days. We have tried to garden here for thirty years. It has snowed here in June. It has snowed here in July, August and September.

But at last we happily report gardening success--in December, January and February--during a frost-free season that lasts 365 days a year in our solar greenhouse. Here with snowbanks mere feet away we are growing in Laramie our first truly successful garden; the first to be spared the catastrophe of cold. Here in our solar greeenhouse we have converted a patch of Outer Mongolia into Florida where we grow tomatoes, cabbages, beans, squash, broccoli, lettuce, spinach, turnips, beets, cucumbers, peas, potatoes and corn.

A TOTALLY SUN-HEATED GREENHOUSE
We owe the success of our garden to Wyoming's brilliant and dependable winter sun, to exceptional insulative values in the perimeter construction and to the unique and economical wet-earth storage bin. The storage is charged from a roof collector by hot air coursing down a large duct to a distribution manifold, then through 21 three-inch plastic air pipes buried in the wet dirt and terminating in a second manifold that returns the air up into the collector for another warm-up (Figures 1 and 2). The plastic pipes (the heat exchangers) are spaced at one and one-half foot intervals where they run horizontally and at nine-inch intervals where rising vertically. Therefore 58% of the pipe exchangers are one and one-half feet apart and 42% are nine inches apart. We use two storage fans; one for the sunshine cycle when the collector is producing heat and a second fan for the night cycle when heat is taken from storage to warm the greenhouse.

How successful is our system? We have maintained during the entire winter a nightime minimum temperature of 58F in the greenhouse by using only heat taken from this solar storage. Hence,

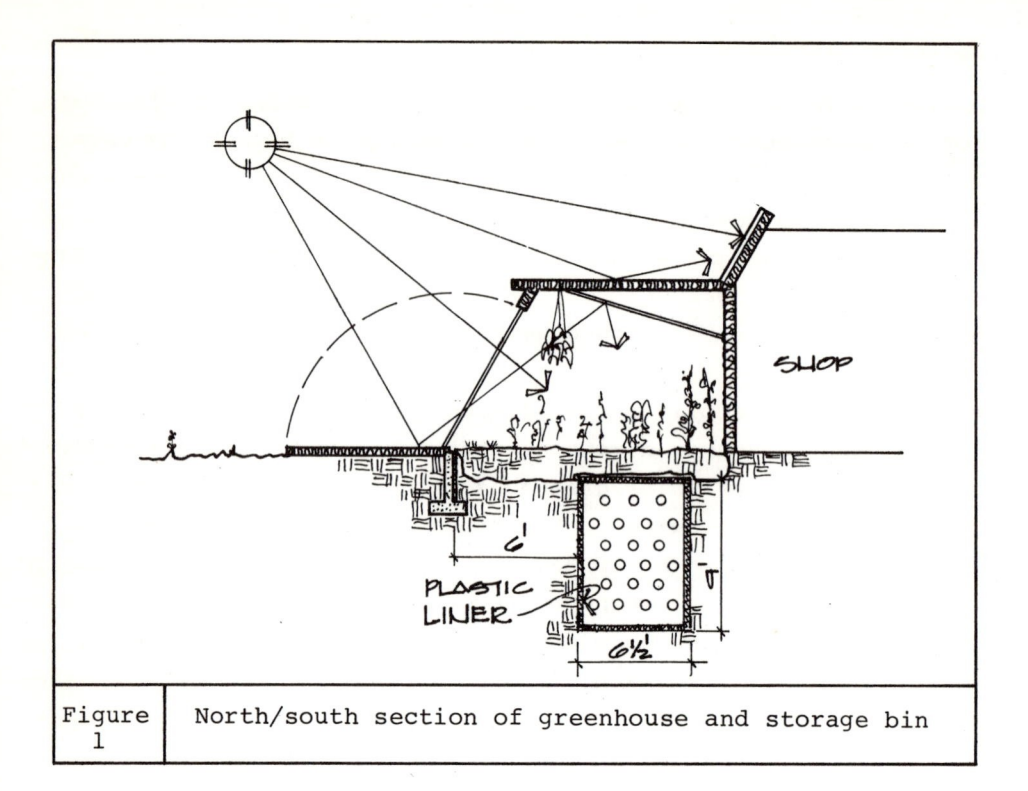

Figure 1. North/south section of greenhouse and storage bin

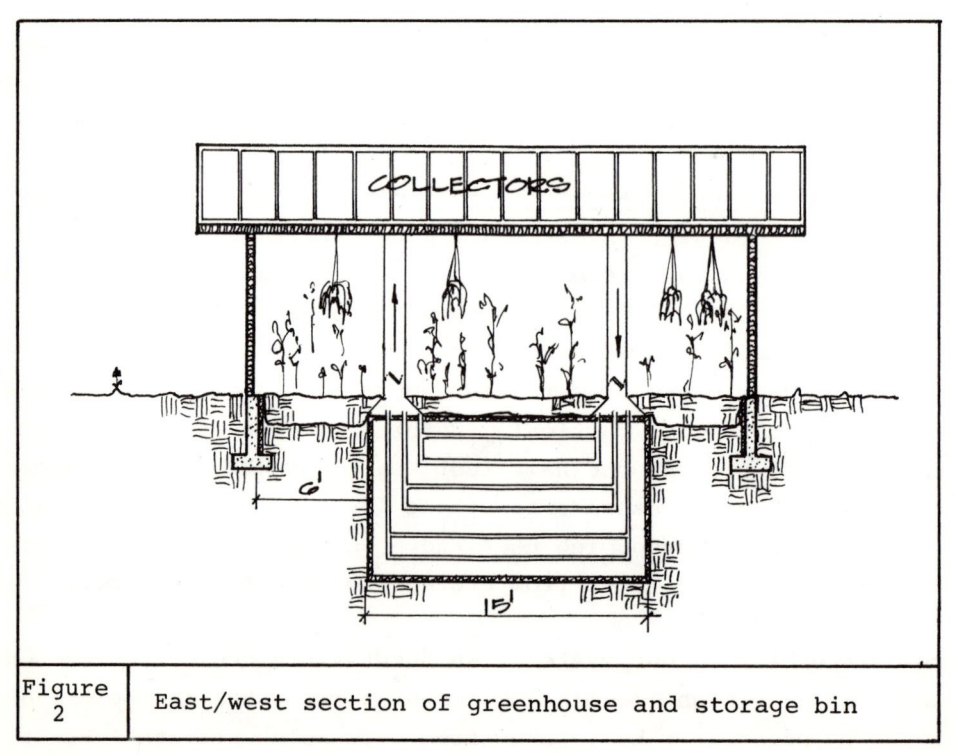

Figure 2. East/west section of greenhouse and storage bin

we label our structure a totally sun-heated greenhouse; and yet it is even more. Not only has the sun met the total heat demands of the greenhouse but has made an additional contribution by warming the one-time neighborhood grocery store--now a large workshop-storage room--to which the greenhouse is attached. On a sunny day the surplus heat build-up in the greenhouse from direct radiation through its glazing is exhausted by fan into the shop, serving nicely to warm this large 40 feet by 28 feet room. On sunless days we do not occupy the shop.

WHY 100% SOLAR HEAT?
We invested our hopes in the concept of cheap, plentiful wet-dirt heat storage because with it we dreamed of achieving total solar heat. We know it is an axiom of the solar economists that total heat from the sun is not yet economically prudent. But we judge it reprehensible to settle for any system of solar heating that will continue to burn up our fossil chemicals. "We cannot afford total solar heat solar heat," say the economists. We cannot afford less than total solar heat, we say. So we must climb Everest. We aspire to the pinnacle of total solar power--because it is there. Because the challenge of the mountain is the very top--not some stopping place part way down.

Anyway, when have Americans--the world's biggest spendthrifts--ever been economically prudent? We know a solar engineer who drives a Winnebago, but who insists that total solar is uneconomical. He advocates 65-75% solar. Would it not make more energy sense, if he converted the purchase price and the cost of operating his Winnebago into payments on 100% solar and then camped out in a pup tent? Since we are habitual and incurable spendthrifts, why not shun a hundred petty squanderings and commit ourselves to one meaningful and positive extravagance--the realization of total heat from the sun? That way we can be the one who shuts off our gas meters--not somone at the wellhead or, worse, someone from Washington. So the Orr family is going for total solar and--though we strive for the most economical means--we do not much care what it costs so long as we can squeeze out enough money to do it. We are going for broke--hopefully by the time we are broke we will have incorporated enough life support systems under one roof, so that we will no longer need much cash.

THE HEAT EXCHANGE SYSTEM

The 21 runs of ABS three-inch plastic pipe buried in the wet-earth bin total 456 linear feet and provide 387 sq. ft. of exchange surface. Why did we use acrylonitrile butadiene styrene (ABS) three-inch pipe? Mainly because it was cheap. And because it was stable to heat. The pipe material was tested in a mass spectrometer and remained stable up to 300°C. A square foot of heat exchange surface from three-inch ABS pipe costs $.44, from four-inch ABS pipe, $.49--so we used the three-inch pipe. Several of our scientist friends, faculty members at the University of Wyoming, were concerned that perhaps the plastic pipe was not a good enough heat conductor to be used as heat exchangers, that we should use something like stainless steel pipe. The cost of steel for this purpose would be prohibitive. A square foot of heat exchange surface from plain steel pipe would cost about $3.00. Anyway, plastic pipe has adequate conductance. The thermal conductance (C) through the 1/8-inch wall of ABS pipe is equal to 9.6 Btu/hr ft^2 F and 2,208 Btu/hr ft^2 F through a 1/8-inch wall of steel. But what is the rush? Heat conducted through either pipe at their respective rates runs smack into nine inches of wet dirt which conducts the heat away at only 1.78 Btu/hr.ft.^{2}F. The plastic pipe's wall conducts heat at about five times the rate that the wet dirt can receive it.

WHY WET DIRT AS A STORAGE MEDIUM?

(1) It is handy, and it is already on the building site.

(2) It has high heat capacity; higher than rocks and lower than water.

> The heat capacity of rocks is 20 Btu/ft^3 F
> The heat capacity of wet dirt is 30-44 Btu/ft^3 F
> The heat capacity of water is 62.4 Btu/ft^3 F

(3) It is cheap. The difference in cost of wet-dirt storage, when compared to other storage mediums, is the cost of its

Solar Architecture   139

   plastic pipe, fittings and plastic liner. Our (878 ft^3)
   (15'x6.5'x9') of wet-dirt storage requires:
       456 ft. of ABS three-inch pipe $171.00
       pipe fittings   35.00
       plastic liner   25.00
                         total $231.00
   Equal water storage would require:
       A 521 ft.^{3} (3,894 gal.) steel
         tank, costing @ $.30/gal. $1,168.00
       Or a concrete tank, costing
         @ $.17/gal.   662.00
       Or 71 gas drums, costing
         $6.50 each (at Laramie)   462.00
   Equal rock storage (1,624 ft.^{3}) or
       82 tons, @ $6.70/ton, costs   549.00

INSULATING THE STORAGE BIN

The tough disadvantage to underground heat storage is that once it is done you cannot make any changes in its size or its peripheral insulation. You should do it right the first time. Thanks to Bill Edmondson we were availed of physicist Jay Shelton's study[11] of underground heat storage in our figuring the heat losses to the ground surrounding our storage facility and how much insulation to install. All of Shelton's results are calculated for a storage region in the shape of a hemisphere with its flat face upwards, because of its mathematical tractability. He states that for other compact shapes, such as our rectangular box storage, "the steady-state heat loss will be approximately the same as from a hemisphere with the same volume, since the temperature distribution would be essentially unchanged a little distance from the storage region."[12] The volume of a hemisphere is $2/3 \pi r^3$. For the sake of calculation we convert our rectangular box-shaped storage of 878 ft^3 to an assumed hemisphere having a radius of 7.49 feet. According to Shelton, the steady-state heat flow from storage to the surrounding ground or the effective U-factor is equal to k/r (conductivity of surrounding soil/radius of hemisphere storage). The conductivity of average soil is 0.5 Btu/hr ft^2 F/ft. So the

U-factor for the ground around our storage is 0.5/7.49 or 0.0668 Btu/hr ft^2 F. This is equivalent to an R-15 thermal resistance. We added three inches of polystyrene ("Drew Foam") with an R-4.35 per inch to the sides of the bin, bringing the total R-factor up to 28. The two feet of wet garden soil over the storage has about an R-2. By adding six inches of Drew Foam polystyrene to the bin's top, we obtain a total of R-28. We did not insulate the bottom. Without it, there is an R-15 in the ground beneath the bin. And since the bottom is only 17% of the total surface area of the bin, we figured that losses would be insignificant. We were wrong.

Does anyone ever add enough insulation? We wish now that we had put six inches of polystyrene in the bottom (costing $87.00) and three more inches around the sides (costing $174.00). We designed for more stored heat than we are getting. We had hoped for enough to heat the greenhouse besides an adjacent apartment in the area we now call the shop. Sufficient heat for the greenhouse and the apartment was to be provided by a heat bin of full capacity based on a heat rise from 80 to 130 F (50 F). Though heat from the collector has topped 150 F the bottom of the bin has never gotten hotter than 90 F and midway up in the bin, only slightly over 100 F. These temperatures have maintained mostly at the hot or heat-entering end of the bin. Hence we came up with only about 20 to 25% of our intended storage capacity. While this is not enough for both the apartment and the greenhouse, there is plenty for the greenhouse. So we are giving all the stored heat to the greenhouse.

We do not believe--as we did--that you need ten days' storage for total solar heat in Laramie. We could have made it this last winter with no more than three or four days' storage. Our estimate now is that five days' storage could get us through any winter in Laramie. So, our recommendation is for a smaller hole in the ground for less storage and thicker insulation around it. Also we think we can do with 36% less pipe. We get about the same temperature drop in the air entering and leaving the bin when we use only the bottom three tiers of exchangers (eleven pipes) or about 64% of the total piping system, as when pressurizing all 21 pipes.

POSTSCRIPT

In recent months we have been drying out the heat bin. We wish to know how dry dirt behaves. Performance data, not yet conclusive, indicates that heat stored in dry dirt is more stable, easier to contain than wet. Already we observe less overnight temperature drop from heat leakage. When the dirt was wet the temperature dropped two to three degrees per night. Dry, it drops a mere 1/2 degree. In theory there should be a loss of conductivity and heat capacity. Performance readings do not show this loss yet. With dry conditions we are recording about the same temperature drop in the air entering and leaving the bin as when the dirt was wet.

A PANACEA FOR WYOMING'S WINTERS

Wyoming has a rather high suicide rate per capita. They say it is because of our long, dreary winters, of the sadness in the constant wind, of the loneliness of the empty and endless miles. But here is a remedy for Wyoming's winters. From out-of-doors, from a cold, dead, grey-white Siberian wasteland you enter upon a warm world of greenness and life. Our solar greenhouse has brought a new joy to winter, a new kind of warmth to our lives; the sun-soaking, humid, hugging warmth that conducts immediately to the bones. We have lived with it for almost two winters. We will never live without it again. The increased humidity cuts down on colds. (In one study, Swiss children had a 50% reduction in colds for pupils in humidified rooms from those in non-humidified rooms.)[13] The sight of green beans growing a few feet from a snowbank, separated by a miraculous Tedlar-bonded membrane (about 1/32 inch thick) finally gives rhyme and reason to Man's long evolutionary haul up from the slime, down from the tree and along the halls of DuPont. For this alone we acquit Science completely for its blunderings. All is forgiven for giving us Tedlar-laminated Filon.

BIOGRAPHICAL SKETCH

Joseph B. Orr is a self-learned solar greenhouse builder and consultant. He has built and tested a wet-dirt storage system for his personal greenhouse.

ENDNOTES

1. Edmondson, William B., 1975, Solterra Home to be built in Laramie, Solar Energy Digest, March.
2. Ibid. 1975, More on the Solterra Home in Laramie, August.
3. Chahroudi, Day, Domebook #2, Bolinas, CA. Pacific Domes, 1971, pp. 120.
4. Edmondson, William B. 1973, The Solterra Home, Solar Energy Digest, November, pp. 1-3.
5. Love, Sam, 1974, Short-loop Living, Equilibrium, July, pp. 43-46.
6. Daniels, Farrington, 1964, Direct Use of the Sun's Energy, Yale University, pp. 97-106.
7. Telkes, Maria, 1955, Solar Heat Storage, in Solar Energy Research, University of Wisconsin Press, pp. 57-62.
8. Shurcliff, W.A., January 15, 1977, Solar Heated Buildings: A Brief Survey, 19 Appleton St., Cambridge, MA.
9. Yanda, Bill and Fisher, Rick, 1976, The Food and Heat Producing Solar Greenhouse, John Muir Publications, pp. 102-103.
10. Jensen, Merle H., editor, "Solar Energy - Fuel and Food" Workshop, April, 1976, The University of Arizona, pp. 177-178.
11. Shelton, Jay, Underground Storage of Heat in Solar Heating Systems, appearing in Solar Energy, Vol. 17, pp. 137-143, Pergamon Press, 1975.
12. Ibid. pp. 141.
13. Conservation Paper Number 43B, Energy Conservation in New Building Design, Effect of Relative Humidity on Health, pp. 248, reported by Rohles, 1975.

AN ATTACHED SOLAR-HEATED GREENHOUSE

Presented by: H. E. "Rip" Van Winkle
Prepared by: H. E. "Rip" Van Winkle, P. Wrenn

INTRODUCTION

A conventional greenhouse, whether is is a free-standing or an attached lean-to structure, consumes energy for winter heating. The following greenhouse is of standard construction, except that it has a heat storage system. This system successfully stores excess daytime solar heat for nightime use, eliminating the need for conventional energy input. Stored solar heat keeps the greenhouse at a minimum temperature of 45 F to 55 F and also provides heat to the adjacent home. This greenhouse, attached to a home east of Boulder, Colorado, (40° north latitude) provides heat and food all winter with no conventional backup heating system.

Figure 1 shows a plan drawing of the 8-1/2 x 20 foot greenhouse foundation which forms a 170 square foot structure nestled in the "U" formed by the front porch and garage of the existing home. The front porch entry is glassed-in to form an airlock area between the greenhouse, the house, and the outdoors. Some of the heartier plants are kept in this air-locked porch.

Figure 2 shows the cross-section (looking east) of the attached greenhouse with a 2400 gallon open concrete water storage tank under the floor. Figure 3 shows a flow diagram of heat storage

and exhange system. Figure 4 shows an isometric view of the
southeast elevation of the greenhouse and solar collector.

GREENHOUSE ENERGY BALANCE

The solar gain is computed by multiplying the average greenhouse
height by the 20 foot greenhouse length. The resulting plane
has an area of 220 square feet at a tilt angle of 45 degrees
from horizontal. The solar gain on this plane is 700 Btus per
day multiplied by 220 square feet, or 154,000 Btus per day.
The condution loss is 1.2 Btus per hour on a glass area of 260
square feet (including the upper shed roof of the greenhouse).
Assuming an inside versus outside temperature differential of
30 F for 16 hours, the daily heat loss is about 150,000 Btus
per day. During daylight hours the fan blows hot air from the
ceiling over the heat exchanger, while water is pumped from the
bottom of the storage tank to the heat exchanger and the bench-
heating pipes. The heat collected by the heat exchanger and
the bench-heater is recirculated to warm the tank water at a

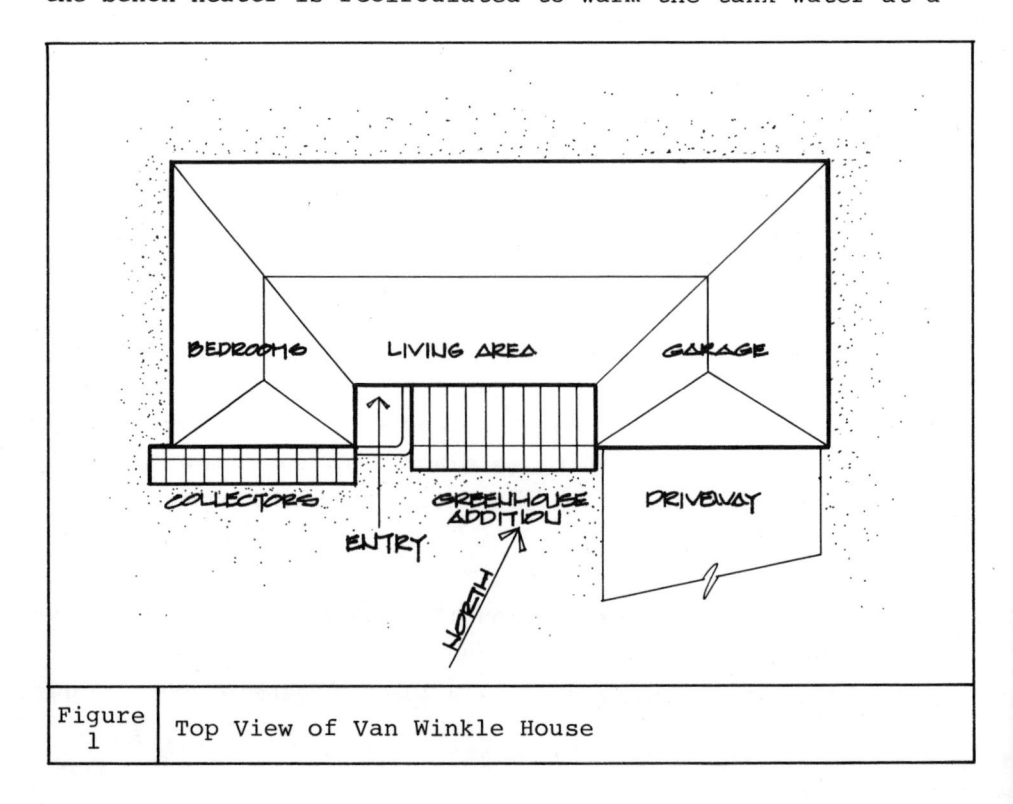

Figure 1. Top View of Van Winkle House

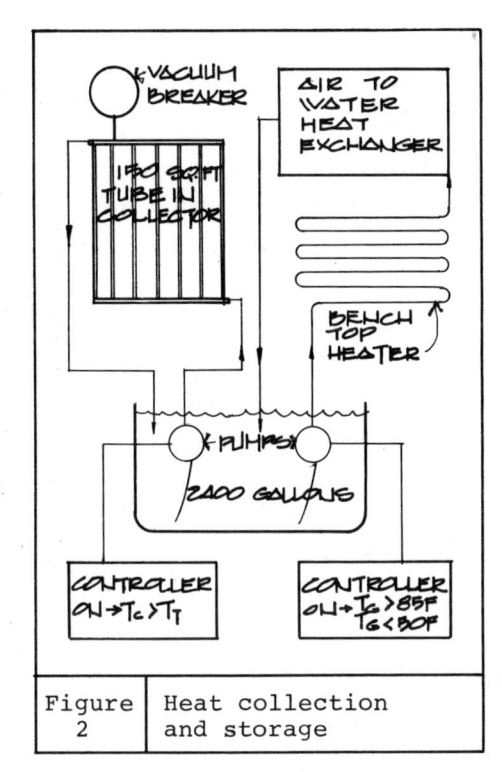

| Figure 2 | Heat collection and storage |

| Figure 3 | Section of greenhouse |

| Figure 4 | Perspective from the Southeast |

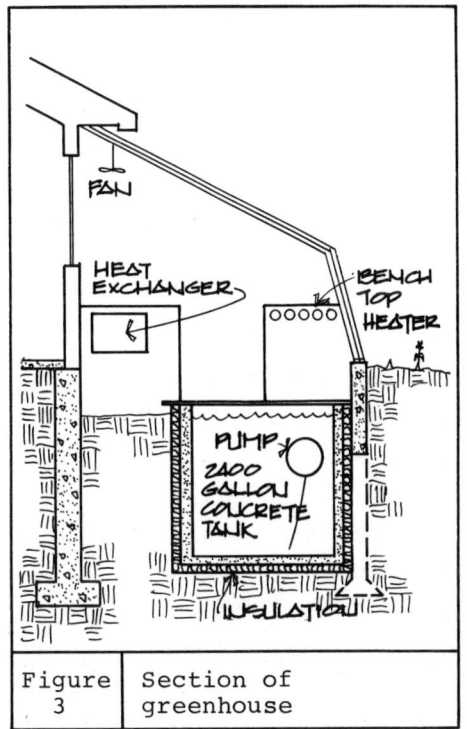

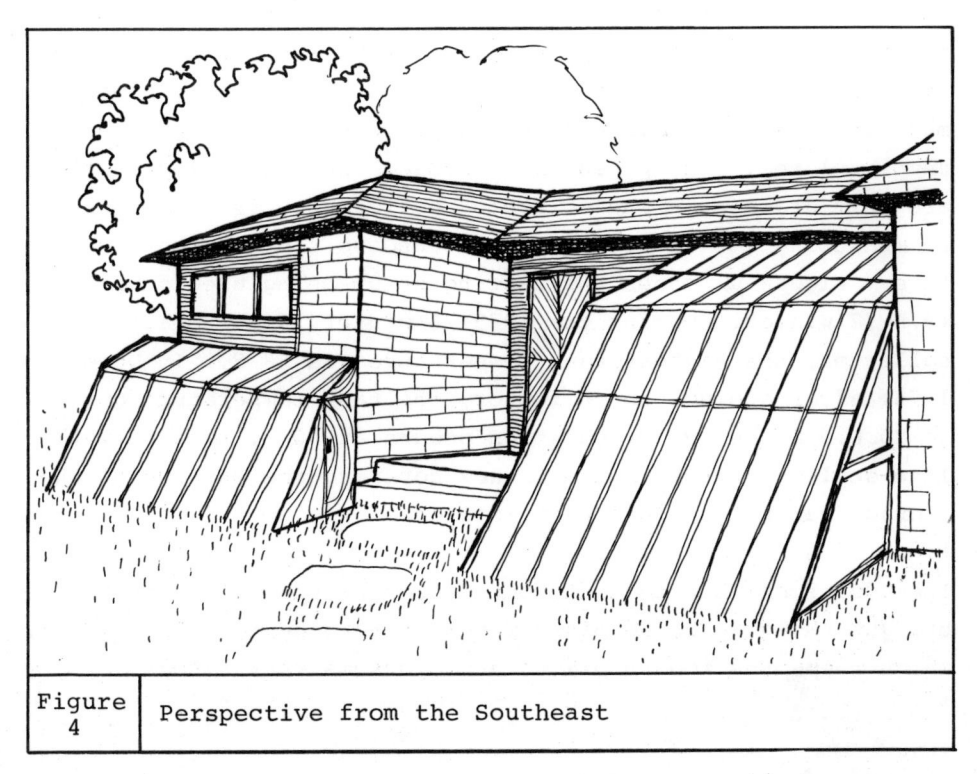

rate of up to 14,000 Btus per hour. This will raise the total storage water temperature about 3 F per day. During nighttime hours the storage tank warms the greenhouse air by conduction and convection. If necessary, the pump and fan can be activated to move heat from storage to the greenhouse at a faster rate.

The large storage volume (2400 gallons) is used to minimize the storage temperature swing and to provide a maximum Delta T between the tank and heat exchangers. The greenhouse temperature swings from a maximum of 90 F to 100 F to a nighttime minimum of 40 F to 50 F. If the tank temperature averages 65 F, the daytime Delta T is 25 F to 35 F and the nighttime differential is 15 F to 25 F.

ACTIVE COLLECTOR AND GREENHOUSE COMBINATION

After using the passive greenhouse for six months we decided to build a matching support structure for an active solar collector on the other side of the front door (see Figure 1). This 150 square foot collector will be used for domestic hot water and hydronic space heat in the winter. The collector is a home made unit using 1/2 inch copper pipe soldered to eight-ounce copper plates. The pipes are arranged and controlled so they are self-draining. The collector performance is being tested by measuring its effect on the greenhouse storage tank temperature. Figure 2 shows the testing setup being used during the 1976-77 winter. During the 1975-76 winter, with no active collector, the highest tank temperature was 68 F. In December 1976, after adding the collector, the tank temperature got up to 90 F. Excessive evaporation occurs at 90 F, so a polyethylene-sealed stryrofoam lid is now covering the surface of the storage water to reduce vaporization.

The additional stored heat contributed by the collector has brought the greenhouse minimum temperature up to 50 F to 55 F, which keeps the tomato plants and blossoms very happy.

RESULTS

The first year of greenhouse operation without the solar collector reduced the home fuel consumption by 25%. As of March 1, 1977, fuel consumption is 7% lower than the previous year without direct solar collector assistance to the house-heating load. The greenhouse is supplying heat to the house by passive convection, as well as conduction through the masonry wall separating the greenhouse from the house. The collector is providing heat to the greenhouse via the storage tank and heat-exchangers.

FUTURE IMPROVEMENTS

1. An insulated cover system for the greenhouse glazing is needed to reduce nightime heat losses and to create more available heat for the house-heating and domestic hot-water systems.

2. Dividers in the storage tank are needed to stratify the storage temperature and provide higher temperatures for domestic hot-water heating and space-heating for the house.

3. A heat pump could be installed to reduce the greenhouse diurnal temperature swing. The heat pump would reduce the daily greenhouse temperature by heating the tank at a faster rate during the daylight hours. This would also provide more heat for nightime use in the greenhouse.

BIOGRAPHICAL SKETCH

H.E. "Rip" Van Winkle is an electrical engineer for IBM and a part-time solar consultant in Boulder, Colorado.

SOLAR COLLECTOR SIZING

Presented and prepared by: Herman G. Barkmann

In a technology as relatively new as the use of solar energy for space and/or domestic water heating, it is not too surprising that there are many ways to determine the area of collectors. Most methods have been derived from earlier studies by Löf and Tybout concerning the economic use of solar energy. Based on several geographic areas, these early studies determined that the system which gave the greatest economic advantage for solar space heating was one in which the solar energy contributed approximately 60-70% of the annual energy required for heating. As a result, a designer might arbitrarily pick a percentage (say 60%), calculate the energy required for the annual heating requirement and then calculate the area of collecting surface required to supply 60%. If the physical characteristics of the structure would allow, this would determine the design. Recent computer studies at Los Alamos Scientific Laboratory have given essentially the same results for the percentage of energy to be supplied by solar energy.

Using this procedure generally results in 1/4 to 1/2 square foot of collector for each square foot of floor area for residential applications. However, estimating the collector performance is a problem. Furthermore, the rule of thumb storage parameters (10 pounds of water or 30 pounds of rock per square foot of collector) could compound that problem.

COLLECTOR SIZING

Within the last two years, a new and valuable dimension was added to the selection procedure by the development of the "f" charts by Duffie, Beckman, Klein and others at the University of Wisconsin. By means of these charts for liquid and air, a designer may enter the characteristics of a particular collector and determine the area required for a specific portion of solar energy contribution. The collector efficiency factor, F_r, the loss coefficient, U_L, and the transmittance-absorptance factor, $\tau\alpha$ must be known. By combining these characteristics with a reference temperature of 212 F, average monthly ambient air temperature, the incident radiation, the heating loads and the area in question, one may determine two dimensionless values, X and Y. Using these values as the coordinates on the "f" chart, the designer may determine the portion of energy provided by that area. Figure 1 shows an "f" chart for air collectors. The abscissa X is $\dfrac{AF_R U_L (T_r - T_a)}{L}$ and the ordinate Y is $\dfrac{AF_R (\tau\alpha) S}{L}$.

One well known manufacturer of air collectors has developed worksheets using the "f" chart method and converting it so that a designer may determine either the area required for a percentage of energy or the percentage of energy for a given area (Figure 2). Here, fortunately, all the variables for both the collector and climatic conditions for several locations are included in the sizing curves, so the procedure is quite simple. Also available is a set of worksheets with which one may calculate X and Y coordinates for nearly any climatic area in terms of the given collector's characteristics. With these values, the percentage may be determined by using the "f" chart.

A manufacturer of liquid collectors has a similar set of curves available which incorporate the performance characteristics of the particular collector. Thus, the designer needs only to know the average outside air temperature, the incident radiation and the heating load, and the percentage of the energy load supplied may b

Figure 1: f-chart for solar air heating systems

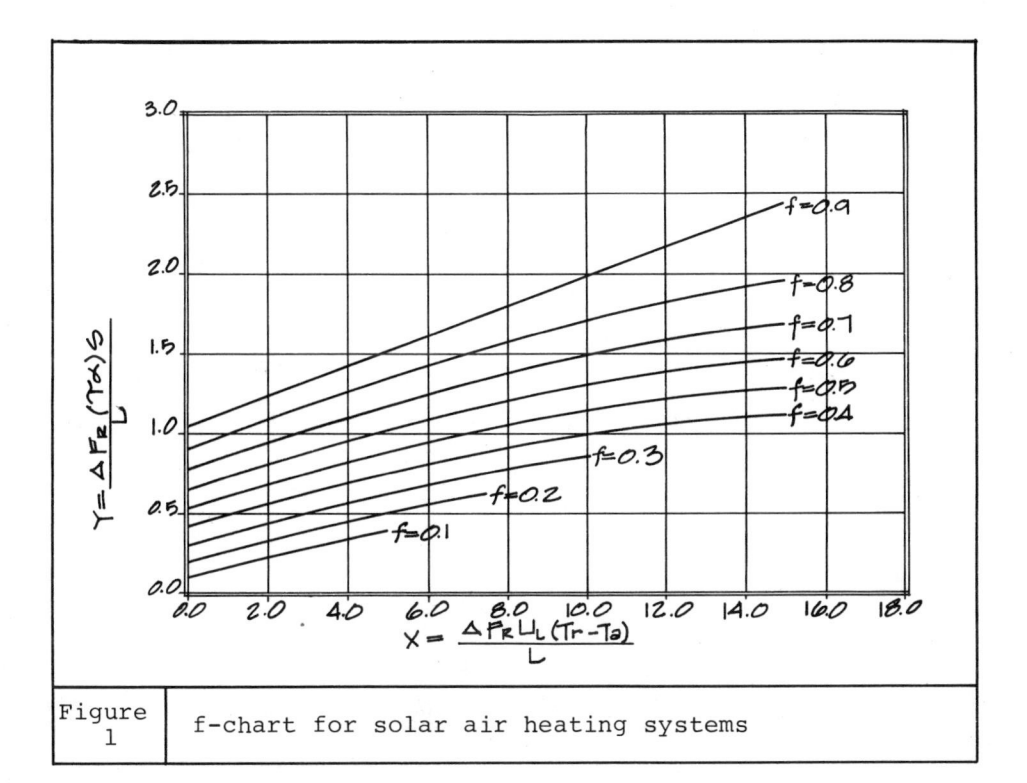

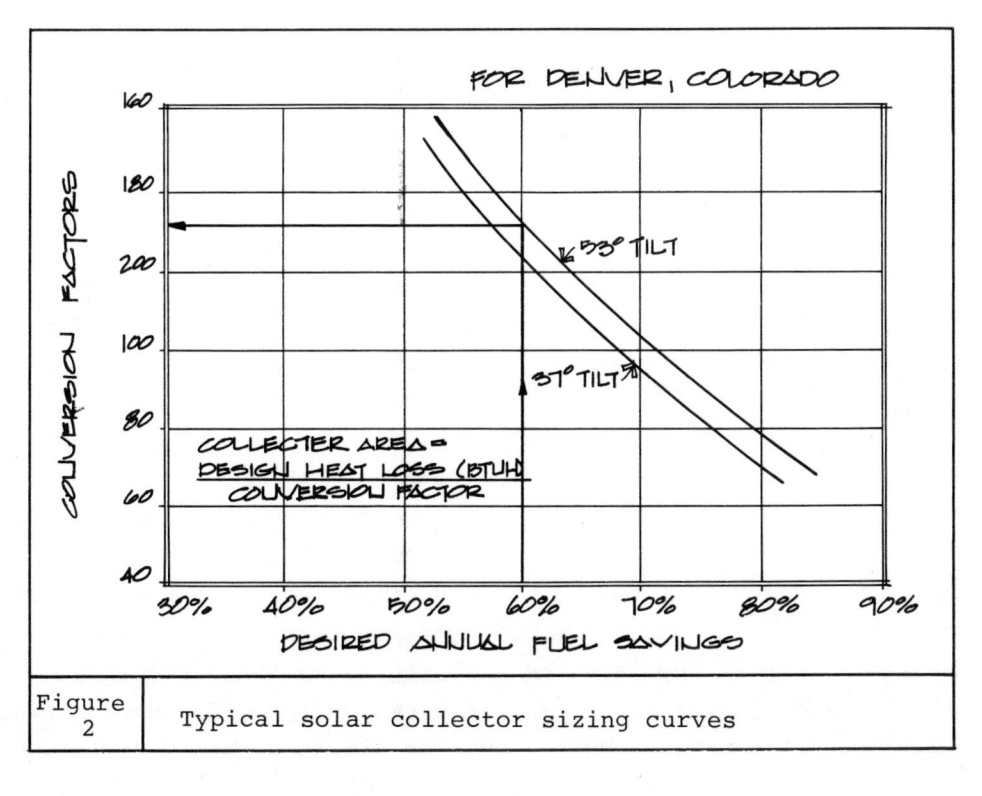

Figure 2: Typical solar collector sizing curves

determined for any collector area. An interesting characteristic of these curves is that one may determine the difference in energy contribution as a function of the available options (Figure 3).

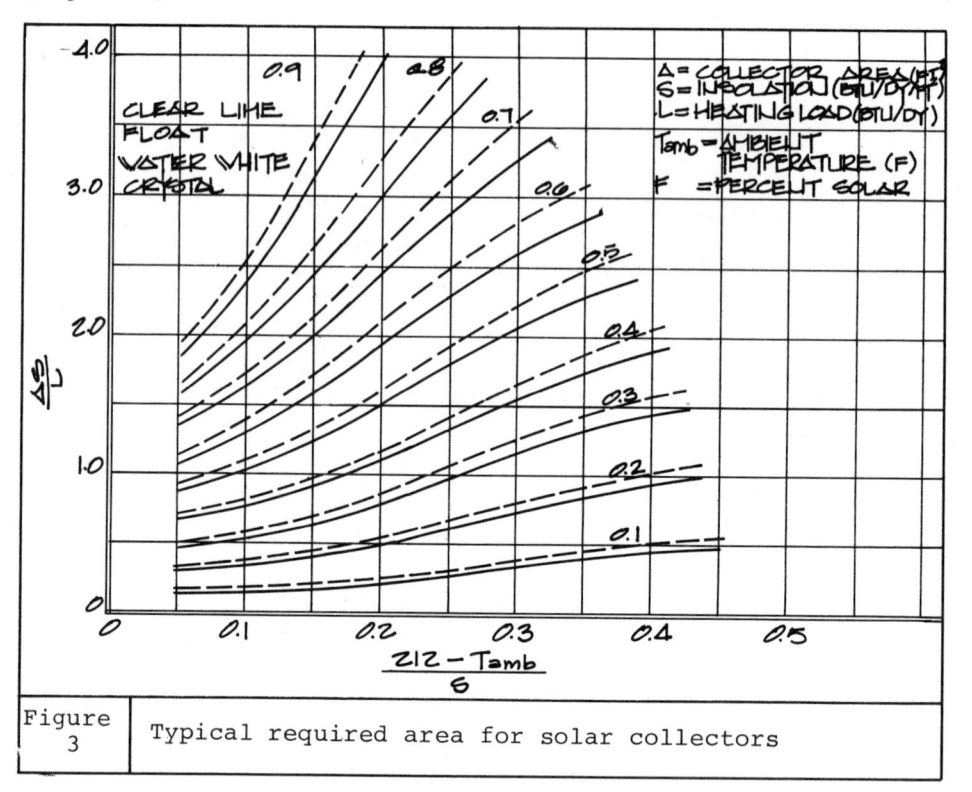

Figure 3. Typical required area for solar collectors

A third manufacturer has made available a program for the Texas Instrument SR 52. This calculator can also determine either the area needed for a predetermined percentage or what percentage a certain area will provide.

The following method of collector sizing for solar heating systems may be used within certain budget and system limitations. Currently, most, if not all, collector manufacturers are publishing the performance of their collectors in the form of slope-intercept curves. Figure 4 shows a comparison of these curves for several different types and manufacturers. The ordinate shows the efficiency, and the abscissa shows the F function. This is a different F; it is determined by the difference between collector temperature and the outside air temperature divided by

Solar Architecture   153

the insolation, I. The efficiency of a collector is of interest
for comparison under a given set of conditions, but the most
value of these curves is the energy output and cost. The important thing we must work for in the solar energy endeavor is how
we can get the most energy for the dollar investment.

A method was developed in an attempt to determine from information available which collector would give the most energy for
the money invested for any given operating condition. Possibly
the most important variable in determining effectiveness of a
collector is the temperature at which the collector is operating.
If one can establish operating conditions and determine the
energy output of any collector under those same set of conditions, a true comparison can be made. If this energy output is
divided by the cost of the collector, a true comparison of Btu
per dollar can be made.

For heating, this method requires that the designer establish the
hourly temperature profile for an average day of some month
during the heating season, usually during December or January.

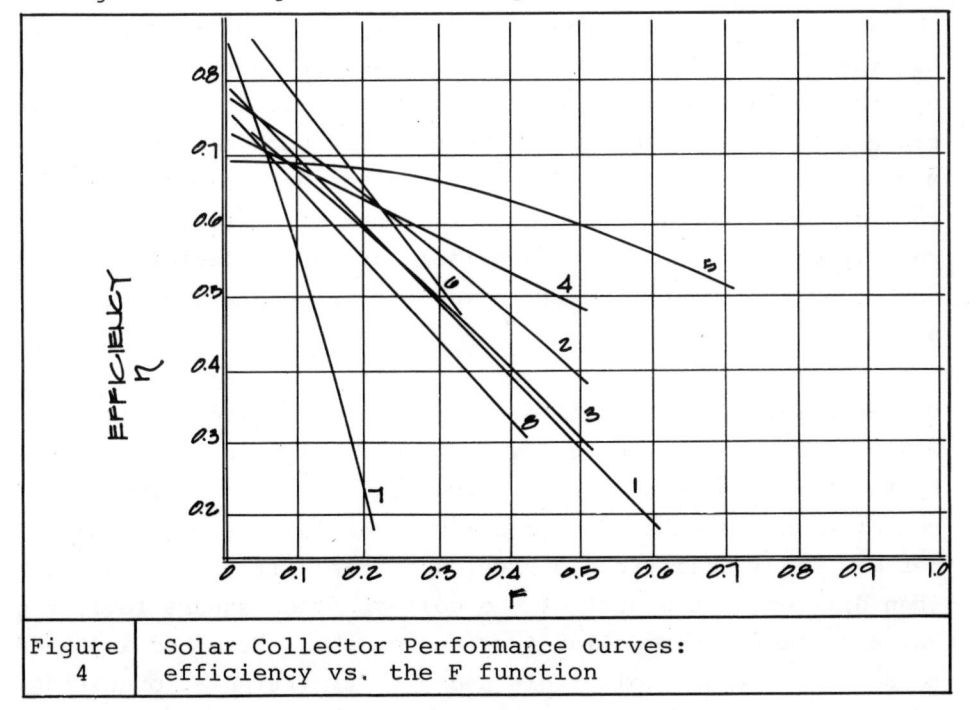

| Figure 4 | Solar Collector Performance Curves: efficiency vs. the F function |

While these profiles are not readily available, they can often be obtained from the Weather Bureau or a local utility. If they are not available for a particular area, the profile from a location similar in climate characteristics can be used.

Also required is the hourly insolation profile for the same month. This can be obtained from insolation tables prepared by ASHRAE and recently extended by the University of Florida, so that the available solar energy can be determined for any latitude at any time. The value obtained should be corrected for sky clearness, altitude and relative humidity. Next it is required that the expected temperature of operation of the collector be determined. This is established by the type of heating system to be used. A hot air system may operate from 110 F to 140 F, a hot-water baseboard at 190 F, a hot-water radiant panel system at 130 F and a water-to-air heat pump at 80 F.

With this data, it is possible by means of the slope intercept curves (Figure 4) to determine the daily output for any collector and compare it against any other under precisely the same operating conditions. The abscissa of the curve is the collector inlet temperature minus the outside air temperature, divided by the insolation at the surface of the collector. This value can now be determined and is the same for each collector. Moving vertically from this value to the performance curve for the particular collector, the efficiency of that collector can be determined. The energy available at the collector face can now be multiplied by the efficiency to calculate the energy absorbed.

By determining the "F" value for each time period available, whether it be hourly or every 15 minutes, the energy absorbed by the collector can then be calculated for that period, and the sum of these calculated values gives the daily energy available per day under conditions of 100% sunshine. If this value is then divided by the cost of the collector per square foot, a value of energy per dollar invested can be determined and used to compare various collectors under that particular operating condition.

Figure 5 shows the results of a comparison of four collectors at four different temperatures for the months of January and June. It can be noted that at low inlet temperatures (collector #3), a simple, inexpensive trickle collector provides more Btu's per dollar investment than the next collector by nearly 50%. Although this collector has the lowest output, its cost is so low that it is still the best investment. Then at the higher inlet temperatures, collector #4 gives by far the most energy for the dollar. This is an expensive focusing collector costing much more than any of the others analyzed, but its higher output at these temperatures makes it the best investment.

This comparison can be characterized by the shape of various performance curves shown in Figure 4. At low "F" values or low temperature, the curves all have essentially the same value. However, as the collector inlet temperature increases, resulting in a higher "F" value, the slopes of the various curves come into effect. This causes an increasing difference in efficiency, which increases the difference in energy output. It can be noted that the comparison of collectors follows the same pattern for winter or summer.

This is first cost of the collector only and does not include installation cost. A low-cost collector that was difficult to install or had a short component-part lifetime would have to have these characteristics taken into account. Here simplicity and track record are most important.

Also, this explanation only shows the way to choose a collector. This needs to be done only for the desired operating temperature, although it can be of interest to see the difference of output for various temperatures.

Once a collector is chosen, the next step is to start the overall system design by first calculating the energy usage for each month. This can be done by determining the average outside air

January

time	ta	I Btu/ft²	$t_i = 100F$ 1	2	3	4	$t_i = 140F$ 1	2	3	4	$t_i = 180F$ 1	2	3	4	$t_i = 200F$ 1	2	3	4
8	22	82	-	-	-	30	-	-	-	-	-	-	-	-	-	-	-	-
9	31	183	75	73	62	115	36	32	19	102	52	47	29	82	32	27	-	-
10	34	250	129	129	116	164	91	88	72	154	85	80	60	140	65	59	70	70
11	35½	290	160	161	147	192	124	121	104	184	95	90	70	172	76	70	131	131
12	35½	303	170	172	157	201	134	131	113	193	88	83	64	182	68	63	164	162
13	33½	290	163	164	150	193	126	124	107	185	58	53	35	173	38	33	175	170
14	40	250	134	135	122	165	97	94	79	156	4	1	-	142	-	-	165	159
15	40	183	84	82	72	117	45	41	29	105	-	-	-	87	-	-	134	123
16	39	82	3	2	-	40	-	-	-	12	-	-	-	-	-	-	75	58
Btu/			918	918	726	1217	653	631	523	1091	382	354	258	978	279	252	914	872
Btu/$			64	83	112	72	46	57	80	64	27	32	40	58	20	23	54	51

June

time	ta	I Btu/ft²	$t_i = 100F$ 1	2	3	4	$t_i = 140F$ 1	2	3	4	$t_i = 180F$ 1	2	3	4	$t_i = 200F$ 1	2	3	4
6	55	15	-	-	-	-	-	-	-	-	-	-	-	-	-	-	-	-
7	60	45	-	-	-	-	-	-	-	-	-	-	-	-	-	-	-	-
8	65	113	53	53	47	73	14	12	4	59	-	-	-	36	-	-	-	-
9	69	170	100	102	94	114	63	61	50	105	23	20	7	90	2	-	20	20
10	71	215	135	139	130	145	100	98	86	138	61	57	43	127	41	37	80	80
11	74	244	159	165	154	165	124	124	111	160	87	83	68	150	67	63	119	118
12	75	253	167	173	162	171	132	132	119	166	95	91	76	157	75	71	143	140
13	76	244	161	167	157	165	126	126	113	160	89	85	70	150	69	65	150	146
14	76	215	140	144	135	145	104	103	92	139	66	62	49	128	46	42	144	137
15	76	170	106	109	101	114	70	68	58	107	31	27	15	93	10	7	121	112
16	75	113	63	63	56	75	25	22	14	64	-	-	-	43	-	-	84	70
17	74	45	10	9	6	25	-	-	-	-	-	-	-	-	-	-	28	5
18	72	15	-	-	-	-	-	-	-	-	-	-	-	-	-	-	-	-
Btu/			1094	1124	1042	1210	758	746	647	1098	452	405	328	974	310	285	889	828
Btu/$			77	102	160	71	53	68	100	65	32	38	50	57	22	26	52	49

Figure 5. Solar collector comparison for 35°N latitude and 45 degree tilt angle

Solar Architecture

temperature for each month of the heating season and using this to determine the heat load for the particular month. If domestic water is to be included, it is then added to the heating load for a total thermal energy requirement.

The next step is to calculate the solar energy available each month for one square foot of collector surface. This can be done in the same way that the comparison was done but for the chosen collector only and from the same weather and insolation tables mentioned earlier. The monthly solar energy must be corrected at this time for the availability of sunshine. The National Weather Bureau has percent of sunshine data for its recording areas. Here again one may have to estimate data by using recorded information from the nearest weather station. Once the above information is gathered and calculations made, the amount of solar contribution for the season for different areas of collectors can be made.

Figure 6 shows a typical calculation. This was made for a design of a project in Colorado Springs and shows the results

Month	Ave. Daily Heat Load Btu/day×10⁻³ (0.67)*	Domestic Hot Water Heat Load Btu/day×10³	Collector Output Btu/sq ft	Sunshine %	Collector Output 1152 sq ft Btu/day×10⁻³	Total Load Btu/day×10⁻³	Solar Usage %*	Total Load Btu/day×10⁻³	Solar Usage %
Jan	927	334	928	74	791	1261	63	1718	46
Feb	884		1144	73	952	1218	78	1654	58
Mar	521		1365	73	1148	855	100	1112	100
Apr	475		1347	72	1117	809	100	1043	100
May	206		709	71	578	540	100	641	90
Jun	-		727	78	653	334	100	334	100
Jul	-		871	77	773	334	100	334	100
Aug	-		800	77	710	334	100	334	100
Sep	112		750	79	683	446	100	501	100
Oct	341		1160	77	1029	675	100	842	100
Nov	536		1049	73	882	870	100	1133	78
Dec	856		820	72	680	1190	57	1612	42

*Energy required above that of heat pump

Annual Solar 91.5%   84.5%
Heating Season
Solar   85 %   75 %

| Figure 6 | Typical calculations of percent solar usage for a Colorado Springs residence |

obtained for one area of collector only. In the actual design, the table was extended several columns to the right for various collector areas considered.

Once one has determined the estimated contribution of solar energy for several different collector systems, the last step is undertaken. That is, to determine the best investment. In the first comparison, only the cost of the collector was analyzed, which is only part of the total cost. Now, the accompanying hardware and its investment must be taken into consideration. What is usually most important to the owner or client is the total annual cost of any system. Generally, the objective is to come up with the lowest annual cost. This is done by estimating the capital costs for the collector area of each particular system, determining the annual cost of this investment, and adding the cost of auxiliary energy. This value can be expressed in an equation $TAC = (A_c C_c + V_s C_s + E_{qs} + E_{qA})I + E_A C_E$ where:

TAC = Total Annual Cost
A_c = Collector Area
C_c = Collector Cost
V_s = Storage Volume
C_s = Storage Cost
E_{qs} = Solar Equipment Cost
E_{qA} = Auxiliary Equipment Cost - A constant
I = Interest
E_A = Auxiliary Energy
C_E = Cost per Unit of Auxiliary Energy

From the above calculations a family of curves can be devised, as seen in Figure 7. The lowest curve is for today's electrical rates, having an ever-increasing total annual cost as the collector area is increased. Typically, at today's costs, the total annual cost is zero collector area. However, if the rate is doubled or quadrupled, then the lowest total annual cost moves away from zero indicating that a finite collector area is a good and proper investment for such an installation as energy rates increase.

Solar Architecture

In the calculation of real costs of solar systems, great care must be taken to properly estimate the costs of all components. The total solar equipment costs for the example shown in Figures 6 and 7 amounted to about $18.00 a square foot. This three to one ratio is a reasonable one to use.

STORAGE SIZING

Storage sizing is also important, and there is a great diversity in concept about the sizing of storage. Often the sizing is done by so many pounds or gallons of storage per square foot of collector. As seen from the above calculations, it is a bit questionable to use this method with the great diversity in collector output. A satisfactory method is to size a storage, so that the heating load of the building can be offset over a certain period of time, usually an average winter day with a moderate drop in temperature of the storage mass. This temperature drop is usually in the range of 30 to 40 F. Too often storage masses are based on 100 F temperature change, resulting in an undersized storage. A small storage means that the storage temperature will rise considerably during collection thereby

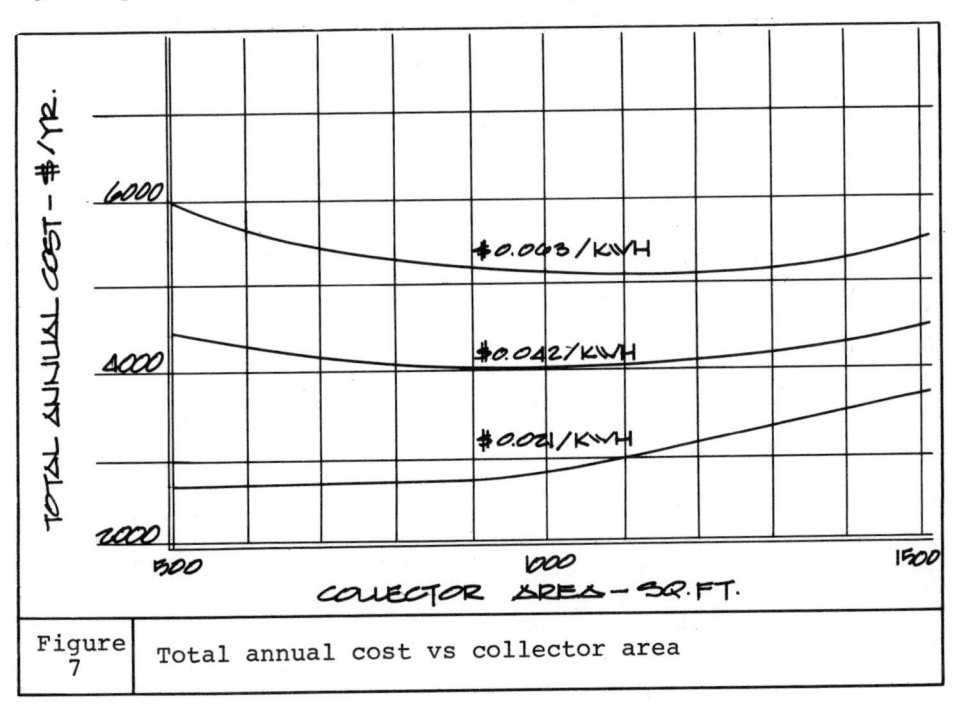

Figure 7. Total annual cost vs collector area

degrading the operation of the collector. If the storage temperature rises considerably, the collector inlet temperature rises, the "F" increases and the efficiency goes down. Column #1 shows the daily output for the same collector with a storage of 10 pounds of water or with 30 pounds of water per square foot. The smaller value ended up with a daily output some 15% less than the large storage.

This problem is improved considerably in storage systems enjoying the advantage of stratification. Rock beds have this beneficial characteristic and here the storage temperature is dependent upon the temperature rise through the collector.

These calculations are worthwhile and can be fun. With the new programmable hand calculators, it is possible to take a lot of the repetition out of the work. These calculations indicate some of many methods used to calculate sizes of solar systems.

BIOGRAPHICAL SKETCH

Herman G. Barkmann is a mechanical engineer in Santa Fe, New Mexico and is involved with the design of a number of solar systems.

RESIDENTIAL APPLICATIONS OF HYBRID SOLAR THERMAL/SOLAR PHOTOVOLTAIC ENERGY CONVERSION TECHNOLOGY

Presented by: Joel DuBow
Prepared by: Joel DuBow, Phillip Risch

SYSTEMS CONSIDERATION

Neither solar thermal energy generation nor solar electrical energy generation alone presently offer viable prospects for residential energy self-sufficiency. As the cost of electrical power increases, and time-of-day pricing by utilities becomes a reality, the combination of both electrical and thermal power from the sun will become an increasingly attractive option for solar homes. Indeed, the most cost-effective solar energy option in the intermediate term could well become hybrid solar thermal/solar photovoltaic energy conversion systems operating under moderate sunlight concentration. For example, the costs of support structure and installation could be shared by both the solar electrical and solar thermal systems. The basis of the hybrid approach is shown in Figure 1. This figure shows the overall system efficiency as a function of fluid temperature. Although the value is given at high intensity, similar trends occur at lower concentration ratios. As the fluid temperature increases, the efficiency of solar photovoltaic cells (N_{cell}) decreases and the efficiency of the thermal system ($N_{thermal}$) increases. However, the overall system efficiency is an increasing function of temperature.

Recent estimates indicate that the cost of solar cells represents about one-third the cost of a solar electrical power system. The remainder of the cost is allocated to interconnections and mounting hardware. In a hybrid system, the solar thermal system could provide the structure for the solar photovoltaic system. In this manner, the cost of electrical energy delivered can be significantly reduced below that of an electrical stand-alone system.

In addition, the cost of solar cells is falling significantly. Figure 2 shows the cost of solar cells as a function of time for both Silicon and Gallium Arsenide systems. Silicon solar cell prices have fallen at a rate of 23% every time the total number of cells produced has doubled. Similar trends may be expected for Gallium Arsenide cells. Gallium Arsenide solar cells represent a high-cost option that will be used only under high sunlight concentration conditions.

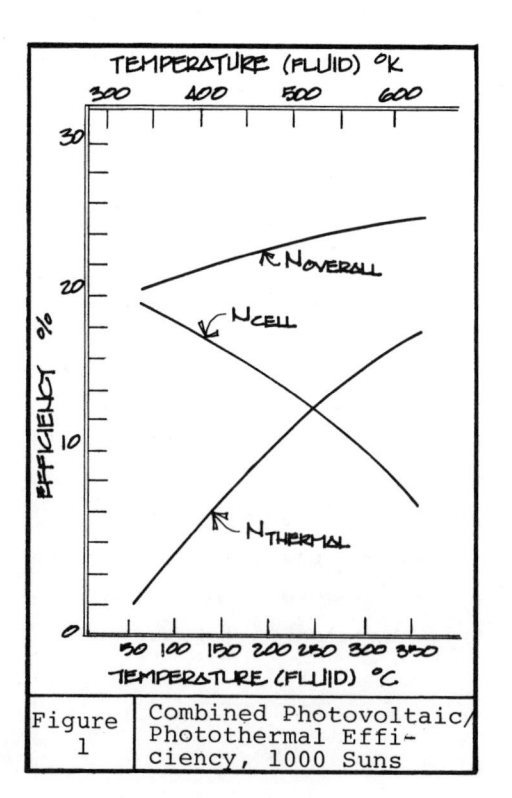

Figure 1: Combined Photovoltaic/Photothermal Efficiency, 1000 Suns

Solar photovoltaic systems may be used in a number of configurations. Some of these are listed in Figure 3. The configuration of a particular system will depend upon many locally determined parameters, including the cost, availability and servicing of the solar system, and institutional concerns such as interest rates and utility escalation rates.

For example, the most cost-effective system at present would utilize little or no electrical energy storage.

Solar Architecture 163

However, that situation may change in the future. In addition, the utilization of available solar energy is reduced without some form of storage. Electronic control using microprocessors could increase the fractional utilization of generated electricity with a minimum of storage. Such controllers are currently under development and will decrease in cost as microprocessor system costs come down. Such control systems could use photovoltaic electricity to control water sprinklers, the solar thermal system, household appliances in anticipation of time-of-day pricing, and telephone diversion systems.

The heart of a hybrid system is the collector. Typical solar electric collector systems are described in Figure 4. The least expensive system today is the flat-plate collector. At present, however, solar cells are more expensive than concentrating optics. A typical parabolic concentrator and rod solar collector is shown in part 2 of Figure 4. Mixtures of these collectors such as a concentrating solar thermal and flat-plate

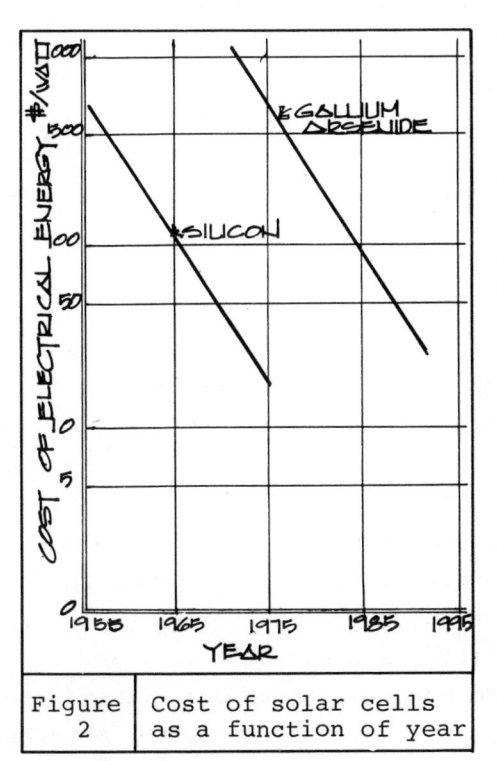

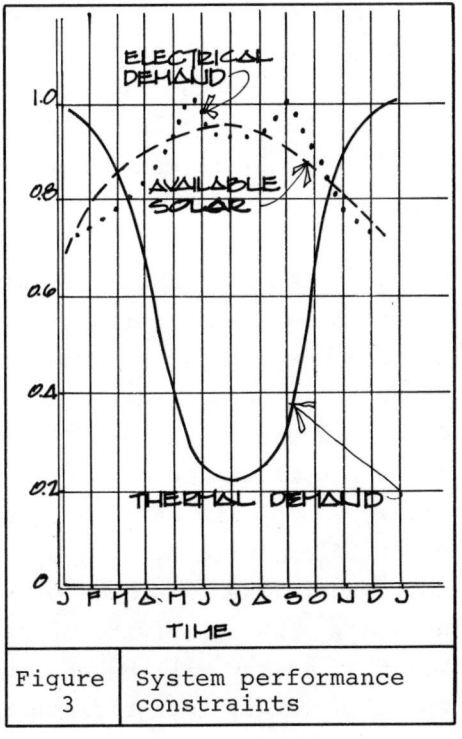

| Figure 2 | Cost of solar cells as a function of year | Figure 3 | System performance constraints |

Solar Cell	Array	Storage	Auxiliary Energy	Interconnection
Silicon	Stationary non-concentrating	Lead-acid battery	Fossil fueled motor-generator	Single isolated residence
Cadmium sulfide	Stationary concentrating	Molten salt advanced battery	Utility grid	Each residence connected to util. grid
Gallium arsenide	Sun-following non-concentrating	Electrolysis cell H_2 storage O_2 storage fuel cell	Fuel cell, purchase H_2	10-1000 residences interconnected
Thin Film cells	Sun-following concentrating	Flywheel	Flywheel motor	1000-plus residences interconnected
		Compressed air	Fossil fueled air compress.	
		Pumped hydro Super conducting magnet	Fossil fueled pump	
		Thermal	Heat Source	
Figure 4	Photovoltaics systems matrix			

solar photovoltaic collectors can also be utilized. The most cost-effective option will depend upon the relative costs of solar cells, concentrating optics, mounting and tracking hardware and bussbar power. If present cost trends continue, then hybrid collectors could be competitive with utility bussbar in a 1990 time frame.

HYBRID SOLAR RESIDENTIAL ENERGY SYSTEMS

Figure 5 is a block diagram of a typical hybrid photovoltaic/ thermal energy system. The roof array will occupy from 25 to 75 square meters. The larger area systems will provide a higher degree of energy self-sufficiency, but at a greater cost and at a lower utilization of available solar energy. Since approximately 1 kW falls on each square meter panel, approximately 150 watts of electricity will be generated for each square meter of collector. The remainder of the solar energy will be generated as heat. The emissivity of Silicon is about 0.3, so a small amount of energy will be radiated back, reducing efficiency somewhat. However, anti-reflection coatings with selective absorbers

such as Indium Tin Oxide, can
significantly reduce the re-
emitted radiation and increase
overall conversion efficiency.

The absorbed energy follows two
cycles: one, a thermal conversion cycle, and the other, an
electrical conversion cycle.
The thermal cycle will consist
of heat exchangers, storage and
coupling to the thermal load of
the residence. The major modification of the hybrid system
will be the utilization of solar electrical energy to power
the pumps or blowers used in the
thermal system.

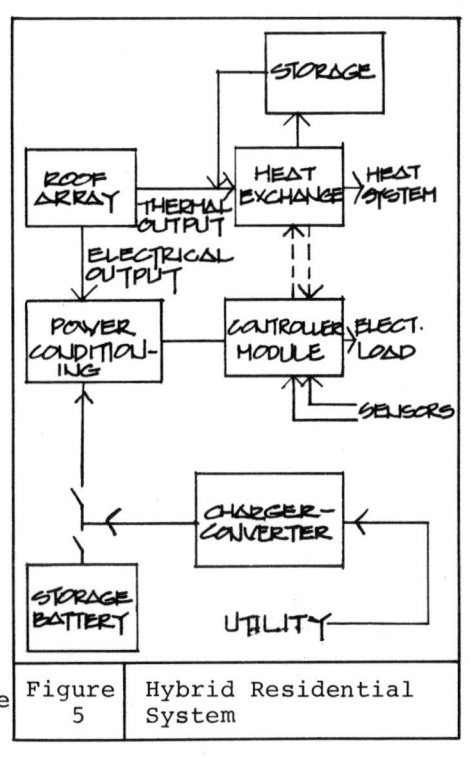

Figure 5. Hybrid Residential System

The electrical system will utilize "power conditioning," which
will convert the DC electricity coming from the solar cells to
AC 110-volt, 60-cycle per second power, which will be mixed and
phase-locked with the electricity coming from the utility. Battery charging will most likely come off the DC before the
power conditioning hardware. Some storage of electricity is necessary in order to attain reasonable utilization fraction of the
available solar energy.

The controller module will control the utilization of solar-de-
rived electrical power and control appliances, sprinklers and
individual room temperature using a preprogrammed alogrithm to
minimize energy costs and maximize the utilization of solar energy.

A block diagram of a system developed at Colorado State University
is given in Figure 6. This system accepted inputs from the homeowner, for example in the form of switches set to control temperature and also from thermostat and other system sensors. All the

Figure 6 System Block Diagram

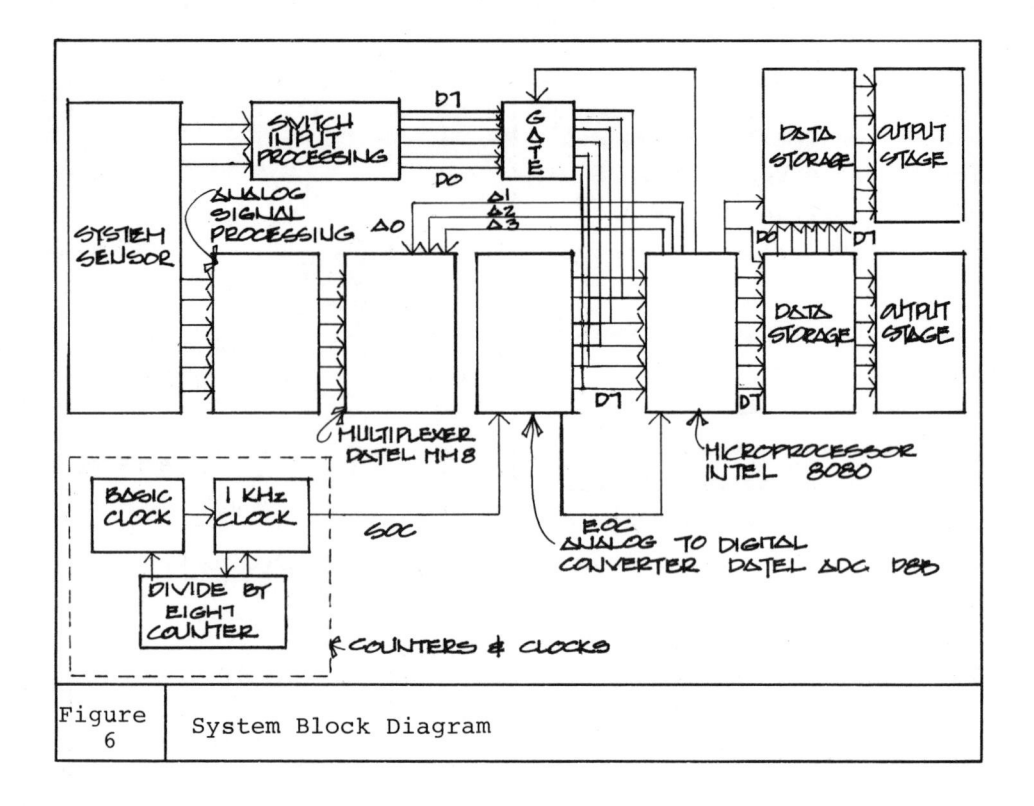

information was converted to digital form, processed in the microprocessor, stored in memory for future reference, and appropriate control signals sent to thermal systems and electromechanical activators for energy utilization and conservation functions.

One problem facing residential hybrid systems, and to a lesser extent, thermal systems, is that the maximum energy is available when it is least needed. At noon the solar thermal system is usually fully charged, and lawn sprinklers are usually not operated. Developing appropriate tasks for the system to perform could utilize energy that would otherwise be wasted. An example might be running a "utility" such as a washing machine by automatic control. A controller could manage this function, as well as manage system-operating temperature. In this manner system temperatures could be elevated when thermal energy is emphasized and lowered when electrical energy is emphasized. By raising the utilization efficiency of the system, a controller will reduce the payback time and hence the "cost" of the solar system.

Solar Architecture

FACTORS IN CHOOSING OR DESIGNING A SYSTEM

Six major elements have to be considered in designing and/or selecting a solar total energy system. The relative cost/performance ratios are site-specific and depend upon both technical and non-technical factors. A key tradeoff will be the desired degree of self-sufficiency versus the utilization efficiency of the system. The six major elements in a solar total energy system are:

Factors of Concern
1. Energy Collection System
    a. Non-concentrating vs. Concentrating, including CPC
    b. Photovoltaic vs. Photovoltaic Thermal or Separate Photovoltaic and Thermal Collectors
    c. Thermal Collector Fluid Temperature vs. Cost and System Efficiency
    d. Fixed, Seasonally Adjusted vs. Tracking Collector Designs
2. Operating Modes
    a. Stand-alone vs. Complete Utility Backup with No Energy Storage
    b. Energy Storage with Utility Off-Peak Battery Charging
    c. All AC or AC/DC Mixed Loads
    d. Utility Owned vs. Community or Individually Owned with Sell-back to the Utility
    e. Microprocessor Load Management vs. Passive Load Management
3. Heating System
    a. Hot Water vs. Air
    b. Heat Pump: Solar Thermal Assisted or Unassisted
    c. Resistance (Electrical Heating)
    d. Conventional Fossil Fuel Heating
4. Air Conditioning
    a. Vapor Compressor
    b. Heat Pump
    c. Thermal Absorption Systems
    d. Evaporative
5. Power Conditioning
    a. Single Inverter vs. Multiple Smaller Units

        b. Parallel Utility Connection to Solar Electrical System
    6. <u>Payback Time</u>
        a. Utility Escalation Rate
        b. Interest Rate
        c. Tax Structure
        d. System Lifetime and Maintenance

At present the limiting factor in system performance and cost is the energy collection system. The heating, air conditioning and power conditioning systems are being independently developed and are under some degree of commercial readiness. Solar energy systems are likely to receive incentives in the form of subsidies and government incentives for the near future at least. A key question for electrical systems is the interaction with the utility system.

One method of minimizing storage requirements and maximizing utilization factors (hence minimizing system costs) is to pipe all excess electrical power back into the utility power grid, in effect running the power meter in reverse. If only a small number of users are using solar electric generating systems, this power into the line will not affect the generating system adversely. However, with many such systems, the resulting electrical load could put sizable transients on the power line and adversely affect system stability. Utility companies are loathe to buy back any power from the consumer, even in small amounts. A compromise allowing electrical energy sellback to the utility grid, could go toward making solar total energy systems viable.

In the next decade, hybrid collectors operating under moderate degrees of concentration (10-40-1) could well prove to be a viable economic option. Solar cells are presently quite expensive ($15,000/kW compared to $500/kW for peaking plants). In addition, optics and tracking hardware for high concentration systems (500-1) are expensive. Therefore, the most cost-effective existing hardware, if full credit can be given for both thermal and electrical energy, could become a moderate concentration hybrid collector. However, considerations of electrical

Solar Architecture

output or thermal output alone could alter these factors. The overall conclusion is quite sensitive to concentrator optics prices, the price of solar cells, and especially the efficiency of solar cells, when operated in a concentrating mode.

MARKETING CONSIDERATIONS IN RESIDENTIAL SOLAR PROPERTIES

More solar homes are now appearing in the residential resale marketplace. Because of their unfamiliarity with such properties, many realtors are faced with new and demanding problems in the marketing procedures. Below are but a few of the considerations that must be dealt with:

Cost. Lengthy and costly engineering coupled to low market penetration and lack of standardization have led to high costs of manufacturing and installing solar systems in private residences. The realtor must be educated, and in turn, educate the prospective purchaser in the economic, performance and reliability factors involved with these systems. Current economic breakeven time frames are averaging about seven years. This figure will of course change as the price of conventional energy increases and that of solar systems decreases. The high cost of solar energy systems has eliminated the under $75,000 purchasers in the Fort Collins marketplace as the average solar home in that area is approximately $130,000. Aside from some notable exceptions, over 97% of the residential purchasers are not able to qualify for one of these solar homes.

Architecture. The requirements of significant collector area and steep-pitched roof orientation has led to the construction of many awkward and unusual home styles. Although it is possible to construct traditional-style homes, most solar homes are contemporary in design. Many objections received in the showing of the solar homes in the Fort Collins area revolve around the home's architecture. As home designers and solar energy systems manufacturers work more closely, hopefully many of these basic-styling problems will be resolved.

Physical Considerations. Since most collectors are placed on the roof and most bedrooms located on the upper level, a major problem has been the noise generated by the solar systems. This is especially pronounced in the hot-water systems. A gurgling or bubbling sound has been heard during system collection cycles. Though not critical, this noise has served as a source of objection to some prospective purchasers. Forced air or heat pump systems tend to have fewer problems' along these lines.

Overview. The residential market in the Fort Collins area, and indeed the entire front range, is an extremely active one. Many prospective puschasers are initially interested in solar homes for both ecological and financial reasons. They soon become discouraged, however, because of some of the reasons mentioned above. As costs come down and other problems are worked out, the residential solar home market will become one of the most active and exciting. There must, however, be a mutual educational program instituted by both the solar energy and realty industries, so that both may work hand-in-hand in bringing this important new aspect of housing to successful fruition.

SUMMARY AND CONCLUSIONS

Hybrid residential solar energy systems provide electricity, heating and cooling to the homeowner. When hybrid systems are coupled to sophisticated electronic controls, high fractional utilization of the available solar insolation, efficient electrical and thermal load management and automatically controlled energy conservation measures will combine to maximize the energy self-sufficiency of individual homeowners.

At present both photovoltaic systems and microprocessor-based controllers are too expensive to be viable economic options for a home buyer. Hybrid solar panels are not yet manufactured, although some are under development. However, intensive research and development efforts are resulting in rapid and significant cost reductions in both technologies. For example, ERDA 1985 cost goals for photovoltaics of 500 per peak kilowatt are expected to be attained and exceeded before 1985. Should present technology trends continue, photovoltaic and hybrid technologies could be economically viable and in a position for economic takeoff.

Activity in hybrid residential systems is presently getting underway under government sponsorship. Both Lincoln Laboratories and Sandia Laboratories are sponsoring programs in photovoltaic and hybrid housing. Colorado State University will have a test bed with hybrid collectors operating during late 1977. Sophisticated microprocessor controls for residences are being developed at Colorado State and elsewhere. Demonstration solar photovoltaic houses are presently being built in Massachusetts. Hybrid houses are being designed and construction will begin in 1978. Houses will be constructed both for experimental use and for residences in communities. Commercial activity could begin in the early 1980's. The 1980's could see a similar development pattern in hybrid and photovoltaic housing that the 1970's witnessed in solar heating and cooling.

BIOGRAPHICAL SKETCH

Joel Dubow is a professor of electrical engineering at Colorado State University in Fort Collins, Colorado.

EUTECTIC SALT AS A SOLAR HEAT STORAGE MEDIUM

Presented and prepared by: Don M. Harvey

A solar-heated building, active or passive, must have a heat storage system for periods when the sun is not shining. During the night and on cloudy days the storage system will be used to provide all or a portion of the heat for the living space. The size of this storage system will depend not only on building requirements, but also on the material used to store the solar-collected heat. The most commonly used method has been the sensible heat method; that is, by increasing the temperature. And the most widely used materials have been water and rock.

The ability of a material to store heat by this method depends on thermal mass. The thermal mass is the density of the material times the specific heat of the material. As an example, consider the heat storage capability of 100 cubic feet of the following three materials: water, rock and a salt compound referred to as Eutectic Salt, as each is raised from 80 F to 90 F, a 10 degree increase.

A cubic foot of water weighs 62.4 pounds. It has a specific heat of 1.0 Btu per pound per degree Fahrenheit. Water has the highest specific heat and is used as the basis to compare the heat holding capability of all substances. The thermal mass of water is 62.4 times 1.0, or 62.4 Btu per cubic foot. And the heat required to raise 100 cubic feet of water from 80 F to 90 F would be: 100 x 62.4 x 10 = 62,400 Btu.

Rock weighs much more than water and it is reasonable to assume that it would hold more heat. However, Basalt rock has a density of 184 pounds per cubic foot and its specific heat is only 0.2 Btu per pound. To get the heat into and out of the rock, crushed rock or gravel is used and the density of rock in this form is only about 90 pounds per cubic foot, which provides a thermal mass of only 18 Btu per cubic foot per degree F. Therefore, the heat storage of 100 cubic feet of crushed rock raised from 80 F to 90 F would be: 100 x 18 x 10 = 18,000 Btu. This is less than one-third of the heat storage capability of water.

Recently considerable research and development has been performed on storing heat by the latent or phase-change method. A phase change means a change in state, like ice to water, or water to steam. Let us consider some substances that will go through a phase change in the temperature ranges we can obtain from a solar collector. There are some salts that melt around 100 F that can be used. One that is inexpensive, easily obtained and available in large quantities is Sodium Sulfate decahydrate, $Na_2SO_4 \cdot 10H_2O$; when mixed with 3% borax, it is referred to as <u>Eutectic</u> <u>Salt</u>. As Eutectic Salt melts at 90 F, it goes through a phase change and it absorbs 108 Btu per pound. It has a density of 91 pounds per cubic foot and a specific heat of 0.84 Btu per pound per degree F. Eutectic Salt has a thermal mass of 76.44 Btu per degree per cubic foot. The heat required to raise 100 cubic feet of Eutectic Salt from 80 F to 90 F would be: 100 x 76.44 x 10 = 76,440 Btu.

But, at 90 F, each pound of Eutectic Salt as it melts increases its heat storage capability 125 times. Therefore, at 90 F, the storage capability for the 100 cubic feet would be: 108 x 91 x 100 = 982,800 Btu/100 cubic feet. This is in addition to the 76,440 Btu's stored in going from 80 F to 90 F. The 100 cubic

Solar Architecture

feet of Eutectic Salt will store 1,059,240 Btu.

A common problem of storing heat in a large mass of salt is stratification when the salt returns from the melted state to the solid state. The solid constituents separate from the liquid, and it requires a large surface-to-volume ratio in order to prevent stratification. Stratification interferes with retrieving the heat from the storage system. Tubes have been tried and failed. A company in Mead, Nebraska (Solar, Inc.) has developed a plastic tray with the proper heat transfer characteristics. Using polyethylene self-stacking trays one-foot square and about 3/4 inch thick, they stack 14 trays per cubic foot and this holds about 50 pounds of the Eutectic Salt. It is permanently sealed in the trays. This method avoids stratification, allows air movement and provides 5,400 Btu per cubic foot.

To put this information in perspective, one may compare the solar heat storage requirements for a 2000 sq ft well-insulated house (with R-20 or better walls and about 12 inches of insulation about the ceiling, weatherstripping, thermopane windows, etc.). A 15 Btu per hour per square foot heat loss in a well-designed house in the coldest days of January is a reasonable approximation. This means the 2000 sq ft house would lose 30,000 Btu per hour and, if the house is heated 18 hours a day, it will require 540,000 Btu per day to maintain a temperature of 70 F. If an active solar system is designed to provide 50% of this heat on the coldest days of the year with a two day thermal storage capacity, 540,000 Btu will be needed. The other 50% can be provided by solar, passively through the windows, by a wood stove or, lastly, by an auxiliary back-up system.

Using the three materials, water, rock and salt, consider water for the heat storage medium for the 540,000 Btu. Flat-plate water collectors normally can heat a properly-sized tank of water to a maximum temperature of 180 F. A fin-tube type baseboard heater will not provide adequate heat to a room, if the water is below 140 F. This means a differential temperature of 40 F. As stated before, water has a thermal mass of 62.4 Btu per cubic foot. To determine the volume of the tank required to provide 540,000 Btu, we divide: 540,000 ÷ 62.4 ÷ 40 = 216 cubic feet. At 7 1/2 gallons per cubic foot, we would need a 1600 gallon tank. This tank would be about 5 feet in diameter and about 12 feet long.

Consider now storing the solar heat in crushed rock. An air collector will heat a rock bed to about 120 F and the hot air from this rock bed is usable in a forced-air system down to about 90 F. Air coming out of a register below 90 F feels cool. This means a storage differential of 30 F. To determine the size of rock bed to store 540,000 Btu, divide: 540,000 ÷ 18 Btu ÷ 30 = 1000 cubic feet. This would require a room 10 ft x 10 ft x 10 ft filled with 100,000 pounds or 50 tons of crushed rock.

Now consider storing the solar heat in Eutectic Salt. Earlier it was found that a pound of Eutectic Salt will provide 108 Btu when it melts and 50 pounds in plastic trays will provide 5,400 Btu per cubic foot. An air collector will heat a storage box of Eutectic Salt trays to 120 F on the coldest days of January. Therefore, the heat storage boxes will provide sensible heat as the temperature in the Eutectic Salt drops from 120 F to 90 F at 0.84 Btu per pound, or at the rate of: 0.84 x 50 x 30 = 1260 Btu per cubic foot. The total heat in a cubic foot of Eutectic Salt trays at 120 F will be: 5400 + 1260 = 6660 Btu per cubic foot.

Dividing 540,000 Btu by 6660, about 80 cubic feet of salt in plastic trays will hold the same solar heat as a 1600 gallon tank of water or 50 tons of crushed rock. A typical solar heat-storage box to hold the 4 ft x 5 ft x 4 ft high stack of salt trays would appear as illustrated in Figure 1. Construction of the heat storage box is simple. It requires not more than one day and can be fabricated by using 2 x 4's and Fiberglas insulation or rigid foam, with 1/2" sheet rock interior and exterior. It is suggested that if this storage box is located in an unheated area, the insulation should be increased to 6".

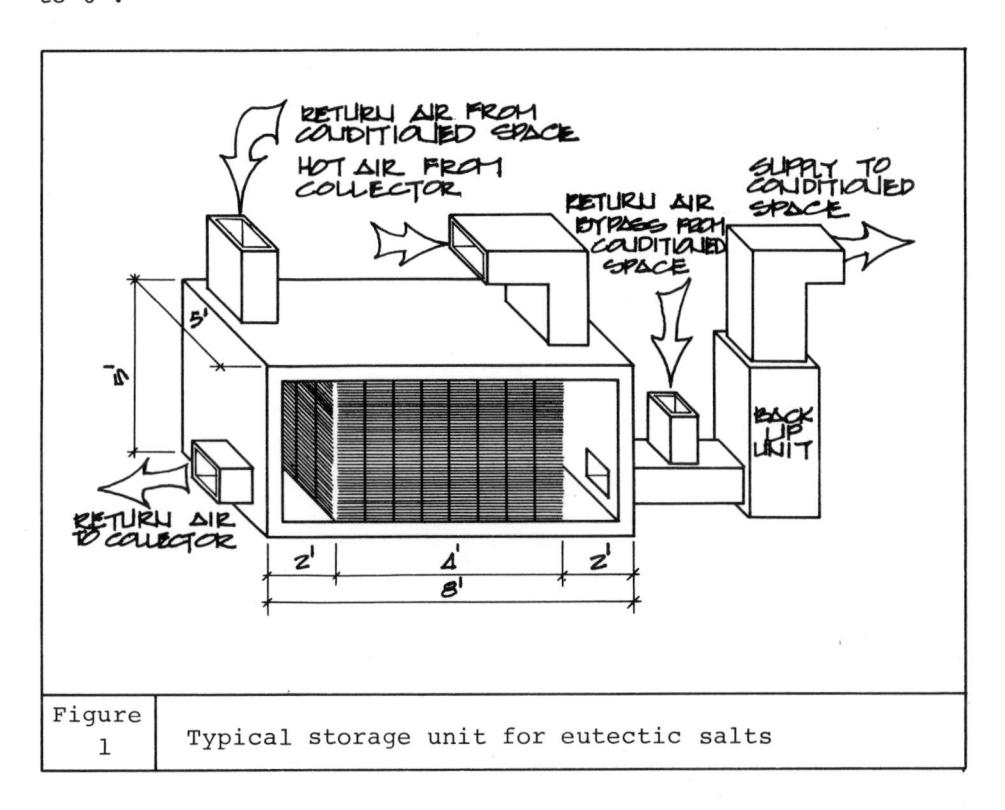

Figure 1. Typical storage unit for eutectic salts

BIOGRAPHICAL SKETCH
Don M. Harvey is a solar engineer with Solar, Inc., Boulder, Colorado.

ARCHITECTURE, THE SUN, AND THE ROARING FORK VALLEY

Presented by: Gregory Franta
Prepared by: Gregory Franta, T. Michael Manchester

The Roaring Fork Valley is nestled in the Colorado Rocky Mountains from Glenwood Springs to Aspen. It has a population of between 25,000 and 35,000 in a 9000 degree-day climate with plenty of high, intense sun, and solar architecture is growing as fast as the environmental awareness of the residents. Through educational programs, like the Aspen Energy Forum, the people are able to make the choice between a home that uses excessive amounts of fossil fuels and one that is more responsive to the environment, making maximum use of the sun's energy.

In a survey of the Valley's solar architecture conducted in May of 1977, the Roaring Fork Resource Center found forty individual solar projects completed and thirty-eight that were either under construction or proposed for construction in the summer of 1977. This survey considered all buildings using the sun's energy either actively, passively or any combination to provide space heating and/or domestic hot water. Active solar heated residences led the collection of solar projects followed by passively heated residences as well as greenhouses, bus stops, workshops, swimming pools, an airport terminal and an apartment complex.

A SURVEY

The following is a sampling of some of the solar buildings that are completed and performing, as well as some that are in or are ready for construction during the 1977 building season.

Craven Residence

The Craven residence (Figure 1) is a contemporary residence designed to utilize solar heating (construction: 1976-77). An estimated 80% of the heating requirements for the Craven residence will be provided by the active solar air heating system.

A separate thermosyphon solar hot water system provides the domestic hot water for the residence. A food and heat producing solar greenhouse is attached to the 2800 square foot residence.

Meadowood Apartments

Energy conservation was one of the major design parameters in

Figure 1 Exterior view of Craven residence (designed by Sundesigns Architects)

Solar Architecture   181

this 36-unit apartment complex (Figure 2). The structure
(construction: 1976-77) is well insulated and each unit has an
air-lock entry. The north, east and west walls only have an
approximate 5% penetration of glazing, while the south wall
has over 50% penetration in glazing for direct solar gain.
The living, dining and kitchen area for each unit has a south-
facing deck. Fortunately, the best view is also to the south.
The total construction cost was approximately $20 per square
foot.

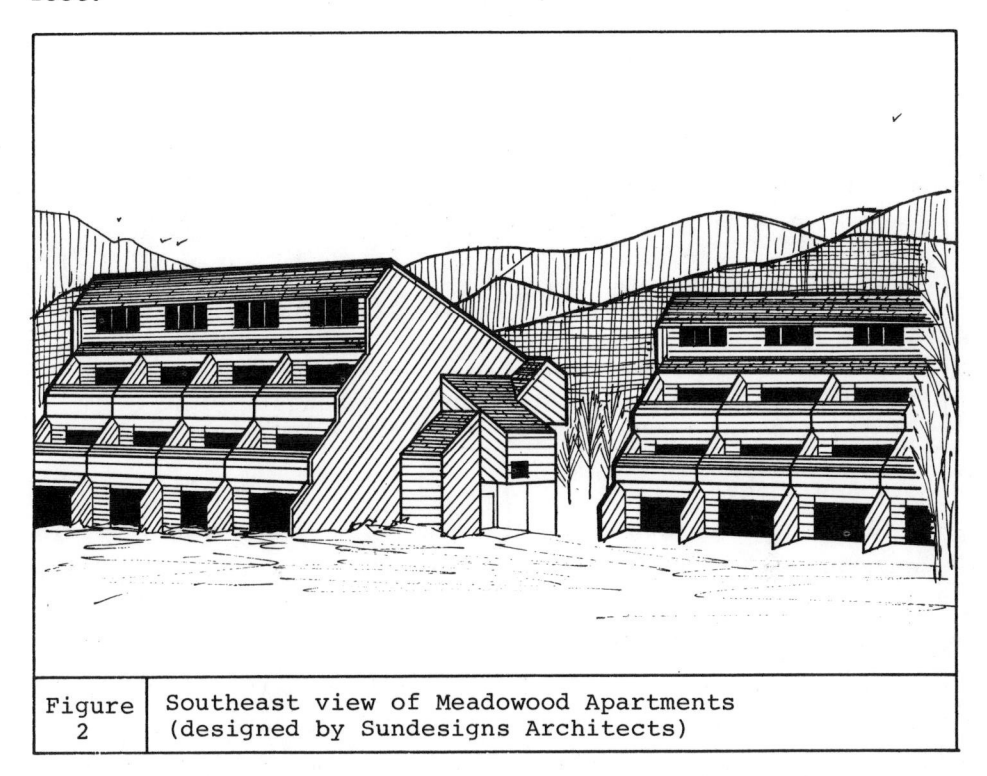

| Figure 2 | Southeast view of Meadowood Apartments (designed by Sundesigns Architects) |

St. Benedict's Monastery

St. Benedict's Monastery (Figure 3) is a 20-year old structure
that is very energy-inefficient and consequently has very
expensive fuel bills.

Presently in design stages is a solar-assisted methane digester
to provide space heating for the monastery in order to reduce
the dependence on fuel oil. The plant has an estimated capacity
of 2,000,000 cubic feet of gas per year with 90 to 95 F heat

provided by flat-plate solar collectors. The methane is to be produced from the manure of 10,000 chickens at the Monastery eggery.

In addition, architectural modifications should reduce the heating requirements by 20% to 25%. A solar system is also planned to provide the domestic hot water.

| Figure 3 | Exterior view of St. Benedicts Monastery (Energy consulting by Sundesigns Architects) |

Shore Residence
The Shore residence (Figure 4) is located in Snowmass, Colorado (construction: 1974). It is a very well insulated 1,500 square foot house tucked into a south-facing hillside. There is 564 square feet of double glazed water collectors with 5,300 gallons of storage in a concrete tank. Distribution is radiant heating from 3/4 inch high temperature polyethylene tube in a concrete floor.

This house also utilizes passive gain through south glazing insulated with beadwall and controlled gain above collectors with insulating covers, reflective on underside for increased gain.

Pitkin County Air Terminal
The Pitkin County Air Terminal (Figure 5) is one of the nation's

Solar Architecture

largest passively solar-heated structures (construction: 1975). The terminal is designed to accommodate a comprehensive transportation center for air, auto and ground mass transportation systems serving Aspen and its contiguous population centers.

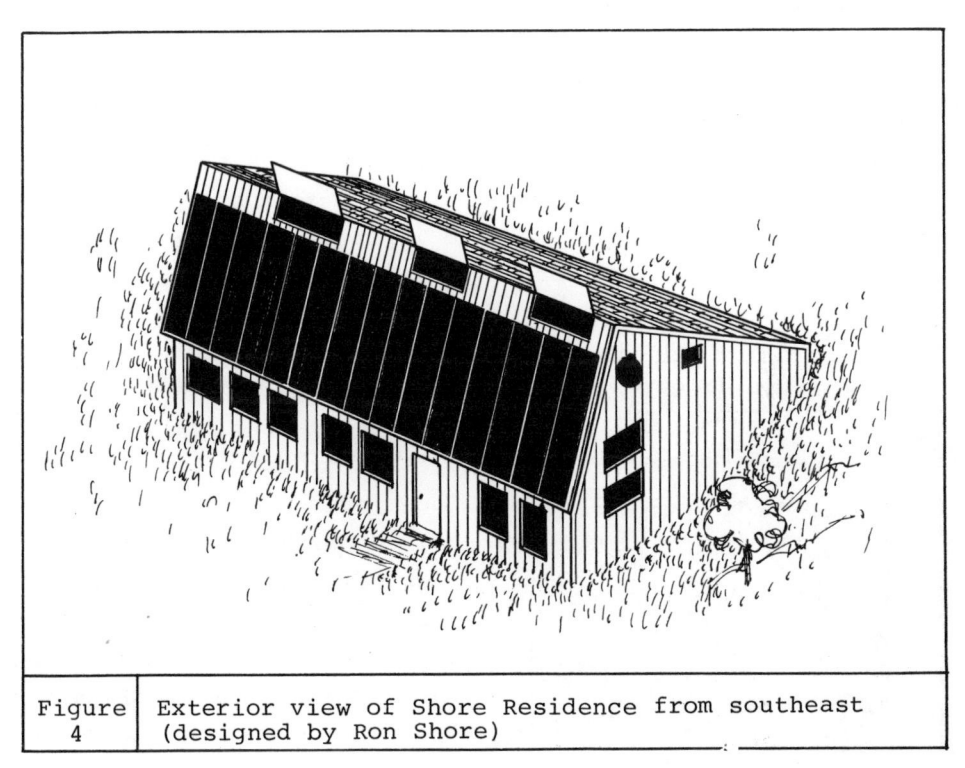

| Figure 4 | Exterior view of Shore Residence from southeast (designed by Ron Shore) |

| Figure 5 | Pitkin County Air Terminal (designed by Copland, Finholm, Hagman and Yaw) |

In addition to the accommodation of specific terminal functions, an overall design objective was resource conservation. The architects designed the building to utilize materials, components and construction techniques that placed a low demand on natural and labor resources for its completion.

The understated architectural character attempts to harmonize with the natural earth forms surrounding the building. To further lower the building profile, as well as to reduce the building heat loss, earth berms are used against all north perimeter walls. Simple and warm interior elements relate the environmental experience of the Terminal to the Aspen character.

Pitkin County Bus Stops

Two bus stops (Figure 6) built by the Pitkin County government contain passive solar heating systems (construction: 1976). The system uses a "Trombe wall" concept with a black thermal-mass wall and a glass glazing. Natural convection introduces the heat into the space during the day while the concrete mass stores heat to be radiated into the space at night. This system has provided a comfortable waiting space for the Valley's bus riders.

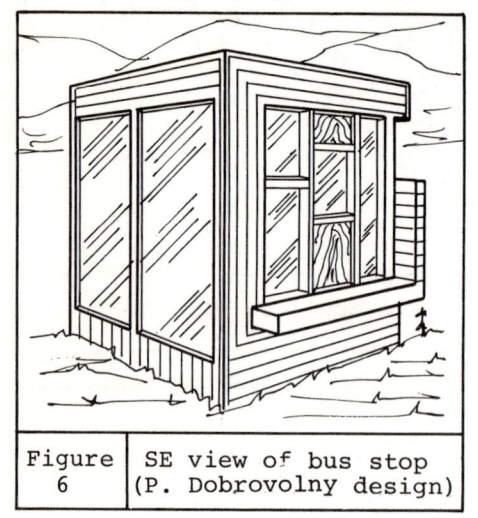

Figure 6 SE view of bus stop (P. Dobrovolny design)

Mollica Residence

The Mollica residence (Figures 7 and 8) is located near Aspen, Colorado at an elevation of 9,500 feet (construction: 1977-78). It is a 1,400 square foot house utilizing a passive-active greenhouse as its collector with thermal mass built into the house's shell. The house is well insulated, has air-lock entrys and is set into the side of a very steep site.

The domestic hot water collectors are integrated into the

Solar Architecture

greenhouse as are two panels of air collectors to boost the air temperature before going to the rock storage. Composting toilets are used in order to conserve water.

Smith-Hite Studio

This building (Figure 9) is an 800 square foot weavers studio with an attached garage and is designed to be 100% independent of fossil fuels for heating and cooling (construction: 1977). Its collection system is an attached greenhouse with transparent solar air collectors and rock storage. There are also five

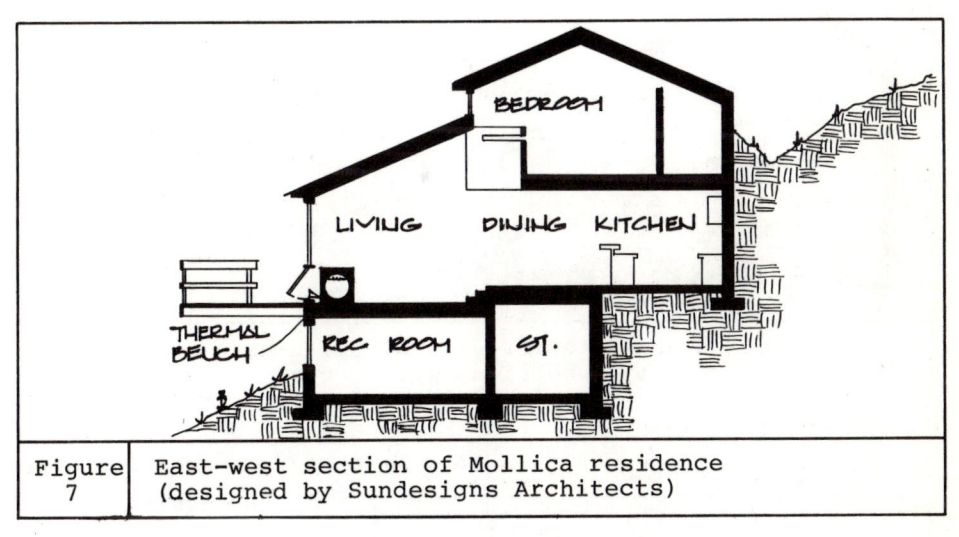

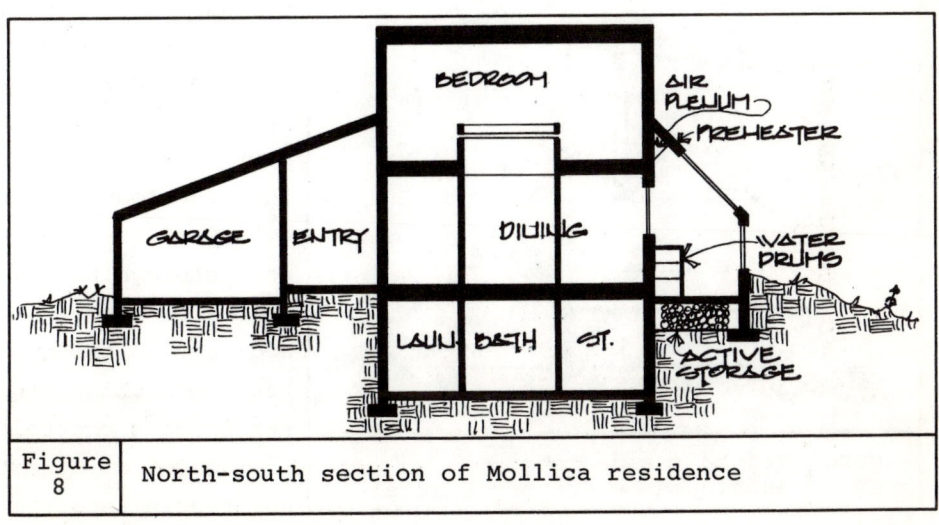

Figure 7: East-west section of Mollica residence (designed by Sundesigns Architects)

Figure 8: North-south section of Mollica residence

skylights and thermal mass in the floor and north wall to passively provide as much heat as possible. A wood-burning stove is used as an auxiliary heat source and for dyes used in weaving.

The domestic hot-water collectors are integrated into the garage along with three more skylights to heat the garage. A composting toilet is also being used to conserve water.

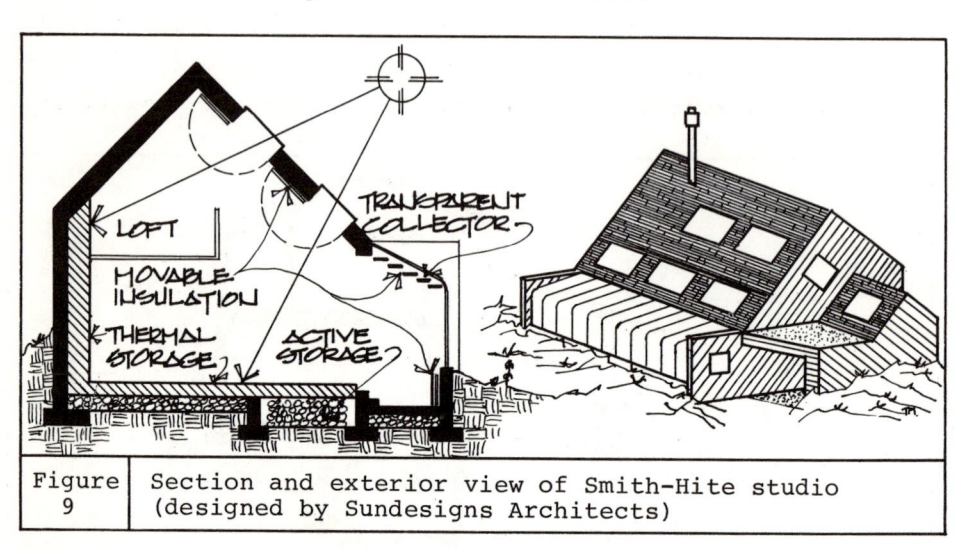

| Figure 9 | Section and exterior view of Smith-Hite studio (designed by Sundesigns Architects) |

Franta Solar Residence

This future residence (Figure 10) for the Aspen area contains 1,000 square feet of living space. It uses passive and active solar systems. The transparent 640 square foot solar air collector, sun-louvers, is connected to a 50-cubic yard storage bin below the first floor. In addition to the active heat collection, the sun-louvers control the amount of direct solar gain into the

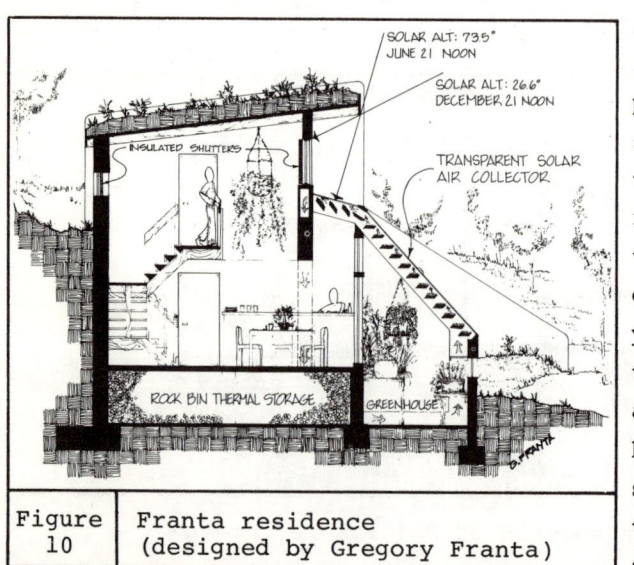

| Figure 10 | Franta residence (designed by Gregory Franta) |

Solar Architecture

heat and food producing greenhouse. It also insulates the greenhouse glazing during the winter nights.

The residence is designed to utilize energy-conserving hardware and appliances. The "Clivus Multrum" organic waste treatment system is utilized to conserve water. Two wood-burning stoves provide the auxiliary heating.

CONCLUSION

The Roaring Fork Valley is becoming increasingly aware of the need for alternatives in architecture at all levels from the owner-builder to the government. Approximately eighty buildings will be purposefully using the sun for heat this winter with more being designed and built every day. It is a natural evolutionary step in the integration of people and environment. It is architecture that is the interface between people and nature.

BIOGRAPHICAL SKETCH

Gregory E. Franta is an architect with Sundesigns Architects and a co-director of the Roaring Fork Resource Center in Aspen, Colorado.

PROJECTS IN TELLURIDE

Presented by: Dean Randle
Prepared by: Dean Randle, Kimble Hobbs, Jack Vickery, Eric Doud, Kathy Noble, Michael Ouellette

TELLURIDE/INTRODUCTION

Telluride, located high and deep within the San Juan Mountains of Southwestern Colorado, has historically been a place where strange and powerful energies have manifested themselves. In recent years, a nebulous group of Designers/Architects have been involved in turning visions into realities. The following is a brief survey of solar architecture in the Telluride area.

Jim Ray's Solar Conservatory

This solar conservatory (Figure 1) was a project for a $60,000 fantasy greenhouse addition, never built. A two-story high sculptural handcrafted iron framework encloses a jungle of plants; waterfalls cascade over the rocks. High in the center is the eagle-claw jacuzzi made of laminated wood, carved and chiseled into the shape of a giant eagle's claw. The jacuzzi has a convective solar hot water system for preheating the water. The heat generated in the greenhouse during the day is mechanically ducted through rock storage along the north wall under the rocks and waterfalls and returned to the space at night when the insulating curtains are raised by ceremonial tassled cords. Have your decadence and be energy conscious too!

Figure 1	Jim Ray's Solar Conservatory

Sunshine Mesa Wind Installation

Five miles west of Telluride, on Sunshine Mesa, Jeff Gerhart of Farmington, New Mexico, designed, fabricated and installed all the components for a wind-generated electrical system for a house. There is no wind generator yet, but the house has sixty 2-volt, low-cycling batteries about the size of car batteries, a rectifier and an inverter, both by Westwind. A propane generator presently charges the batteries; the system provides both 110-volt AC and DC power to the house.

Specie Mesa House

The Specie Mesa house is an integrated-systems house designed and built in late 1976 on Specie Mesa, ten miles west of Telluride. The house is constructed of 10-inch logs locally milled. It combines an active solar hot air space heating system with a wind generated electrical system. The heart of the solar system is an integral roof solar hot air collector designed by Telluride Designworks. The collector cost has been minimized in order to make the system most economically viable,

Figure 2. Specie Mesa House

yet the collector is efficient, durable, easily fabricated and installed, and generally designed for long useful life in the severe climate of mountainous and high desert regions. The collector components include a black absorber duct which fits between 24" on-center roof joists and a double layer fiberglass sandwich panel which is lag-bolted to the tops of the joists to form a weather-proof roof. The absorber duct is a galvanized sheet-steel duct, the top side of which is primed and painted with Nextel velvet black. The duct rests on insulated side supports between the joists. The cover is a double layer of Tedlar-coated Filon, the top layer of which is corrugated. Filon is an acrylic fortified polyester resin reinforced with fiberglass. A redwood frame between the two layers of Filon makes a rigid panel which is lag-bolted to the joists. The backside of the absorber is insulated with fiberglass batts between the joists.

The collectors were fabricated by Telluride Designworks for $3.68/sq ft in materials, $.45/sq ft in labor, and installed for about $1.00 in labor. Because it is an integral roof, the cost of an equivalent conventional roof can offset even more the cost of the collector roof ($1.00/sq ft). Minimizing the cost of the collector is the most important factor in reducing the initial cost of an active solar system, thereby making the system economically viable. More expensive components or processes that would increase the efficiency of the collector only slightly were avoided due to non-justifiable cost-benefit tradeoffs. The collector strives to make minimum use of energy-intensive materials and processes.

The electrical system of the Specie Mesa house has a much larger set of storage batteries: 60 2-volt/580 amp hour batteries, each of which measures 18" wide, 12" deep and 24" high. The batteries cost $1,500 and have a 50-year life expectancy. This spring the batteries supplied DC power to the house for a two-month time period when the propane generator was broken. Jeff Gerhart did the electrical system in this house. The generator (propane, soon to be wind) supplies power through the westwind rectifier to be stored in the batteries. The batteries in turn supply DC current directly to lights, DC outlets and appliances in the house, as well as supplying AC power through the westwind inverter to the AC outlets and appliances. The solar system fan is DC, but the controls are AC, unfortunately requiring constant running of the inverter. (Anyone working on DC controls?) Economic justification of a small scale wind system is near impossible if one lives near available electricity; the Specie Mesa house is two miles from the nearest power lines.

Other design features of the Specie Mesa house include a hot water preheat coil which runs through the rock storage compartment and into a preheat storage tank, and two fireplaces with glass doors each drawing their air for combustion from the outside through adjustable air intake dampers. Wood is the only auxiliary heat source in this house. Future plans include a connected greenhouse and a methane system to make this house an example of relatively individual self-sufficiency.

Solar Architecture

Solar Hotbed/Hillside Hothouse

A way of extending the growing season in cold climates, the solar hotbed is a totally passive structure. Heat generated in the bottom collector/planting bed convects up and into a gravel thermal mass under the top compartment planting bed. Kale and lettuce were planted in early February 1976 and lived through the dead of winter at 8800 feet with no auxiliary heat and no mechanical inputs, just passive solar. It could be expanded into a more complete convective loop passive system in the form of the hillside hothouse.

Solar Victorian

Located in the high "Victorian district" of North Central Telluride, the Luhman house will be the first solar structure in Telluride. The house has energy conserving and passive solar heating and solar heating of domestic hot water. It is projected that the passive solar design aspects of the house will provide 26% of the total annual energy requirement, and the combined passive and active systems will provide 74% of the total annual energy requirement. The initial cost of the solar systems is $7,000 more than the cost of conventional electric heating, but the annual energy savings is $1,134, resulting in a straight payback period of 6.2 years.

The house is characterized by its ability to collect, store and efficiently utilize solar energy. The greater the degree that a structure can collect, distribute and store solar energy by passive or non-mechanical means, the less the structure will have to depend on mechanical active systems. This not only makes the structure more responsive to natural climatic forces and less dependent on external energy inputs, but is inherently more cost effective, because passive aspects of the design can more readily be absorbed into the building cost.

The major form determinants of the house are the passive solarium orientation and the active-collector roof orientation. The house opens itself to the east and south, and turns its back to the north and west. All entries are protected from winter storm winds and are air-locked to prevent thermal loss. There is little glass on the north and west. Porches on the east and south admit winter sun, but shade glass areas from the summer sun.

As a passive design aspect, the solarium is the heart of the house. It collects, distributes and stores energy through natural thermodynamic processes. The floor of the solarium is a one-foot thick paver brick/gravel thermal mass, insulated from the earth, which stores heat from the solar radiation striking its surface and reradiates it back into the space at a later time. Both the floor thermal mass and the fireplace thermal mass act to moderate internal temperature swing because of the time lag between collection and reradiation. Heat can be manually ducted into the cool northwest corners of the house through windows upstairs which open into the solarium. Excess heat in the solarium is mechanically ducted into active rock storage whenever the upper solarium air temperature is hotter than rock storage temperature. The solarium can also be manually vented at the top directly to the outside. The solarium opens directly to the outside, through french doors, to the outside terrace, to invoke natural cooling in the summer time. Existing cottonwood trees on the site will shade the solarium from excessive heat gain.

The solarium glass faces 25° east of south to take advantage of early morning solar radiation, while refusing radiation during the afternoon period. The solarium glass recesses back within the surrounding house mass, utilizing the house mass and the roof overhang to shade the glass in the summer and to protect it from winter storms. The sun stays on the glass a longer period of time during the day in the winter, because it subtends a smaller angle over a given period of time.

The walls and the roof have the economically optimum amount of insulation for this area, six inches in the walls and twelve

Solar Architecture

inches in the roof. Most windows are fixed-glass thermopane with small openers for ventilation. The curtains over the solarium glass, as well as over the rest of the glass, are two-inch thick quilted insulating curtains which seal airtight around the edges. The solarium curtain has a reflective material on the outside to reflect incoming radiation back out during overheated periods. The solarium curtain is an aperture of draperies which manually responds in size and orientation to different patterns of energy.

The fireplace has airtight glass doors and draws its air for combustion from the outdoors, through a manually operated air intake damper. This allows the fireplace to be shut down like a wood stove to burn efficiently, as well as preventing preheated air from going up the chimney. Because infiltration and ventilation is at a minimum, all air circulating within the house passes through an electrostatic filter.

The collector for domestic hot water is a high-efficiency solar hot water collector recessed into the fireplace face. Being within the solarium, it will not experience freezing problems. It is a thermosyphon system, with natural heat convection occurring between the collector and the hot water storage tank. The tank is located at the top of the active rock storage compartment, which keeps the tank as warm as the surrounding rocks. An air passage connects the fireplace heatilator air with the bottom of the water collector absorber cavity, to provide additional water heating potential from waste fireplace heat. An electric immersion heater in the hot water storage tank provides backup heating.

The house has an active solar hot air system for space heating. The collector is of Telluride Designwork's design and covers the southeast and southwest roof planes. The roof planes form two L-shaped solar pockets. This allows a greater latitude of daily collection time, especially in the early morning, as well as utilizing reflected radiation from the adjacent roof plane receiving direct radiation. The reflective porch roofs below the collectors also increase the radiation striking the collectors during the winter low sun.

A thermodynamic simulation of the heat flow characteristics of the house is set up to evaluate the passive and active aspects of the design. Heat loss calculations determine the house heat load. This, added to the hot water heat load, determines the total energy demand of the house. A simulation of the passive heat collection, retention and distribution capabilities of the house is coupled with the sizing of the active aspects of the solar system to determine the total collectible energy by the house. Monthly percentages of the energy demand which can be met by passive solar energy, active solar energy and auxiliary heat can be determined. A life-cycle cost analysis compares the additional initial cost of the solar system to the cost of fuel saved over the lifetime of the system to determine the economic viability.

The Solar Trough

Just beginning construction on Turkey Creek Mesa south of Telluride, the solar trough is a third-cycle HUD solar grant proposal. The house attempts to integrate energy-conserving features (passive, hybrid and active systems) into a balanced whole. The house is recessed into the earth, almost underground; the back wall retains 24 feet of earth. Heat losses are minimized by insulation, airlock entries, a skyshaft with a manually operated insulating panel and insulating louvers, which double as a passive/hybrid heat collection system over the greenhouse. Purely passive aspects of the house include insulated thermal mass floors, solid thermal mass bearing and retaining walls, a hot tub in the center of the greenhouse passive/active rock storage floor, and a storage-to-space heat distribution system which has been eliminated by allowing controllable natural convection with various doors and windows between levels acting to control and direct heat flow. The hybrid aspect of the house is the manner in which the louvers are able to respond to different conditions. The louvers can

Solar Architecture 197

open to let maximum energy into the space, or they can close to shade the greenhouse while acting as a relatively efficient active solar hot air collector, air being mechanically ducted into rock storage, and the louvers will close during periods of heat loss to insulate the glass. We will attempt to automate the louvers to some reasonable degree.

The domestic hot water tank is painted black and set half-way into the greenhouse floor/rock storage bed for preheating. There is a small active solar hot air system in the 45-foot long, 8-foot wide trough. The collector covers standard are 46-inch patio doors spanning the roof joists to get comparative data to the Filon covers. The Clivus dry toilet, rather than being hidden in the cold dark basement, will be painted black behind a glass enclosure to really get that stuff cookin'. A grey water re-use system will be incorporated into the house also.

Once again a performance simulation is set up in an attempt to predict the passive and active shares of the contribution to the thermal balance of the house. 39% of the total heat load should be supplied by the passive/hybrid aspects of the system and 45% should be supplied by the active aspects; a total of 84% supplied by solar. Straight payback period for the active system - 7.9 years; for the active and passive/hybrid systems - 13.7 years. The relatively high-tech nature of the solar trough initially seems to be revealed in the longer payback periods. It will be possible to compare the functional and economic aspects of passive/active, low tech/high tech when more working data comes in on these houses to compare with the predictions.

Lazy-O-Ranch
The four Ouellete brothers and friends are building the most ambitious integrated systems project in the area at Horsefly Mesa, 20 miles west of Telluride. A 7,000 sq ft structure is in its second year of construction. A stonehenge circle of piers surrounds a thirty-foot monolith of a fireplace. Stone-

work was completed last summer (1976). This summer the structure will be enclosed and a hot water space heating collection and distribution system will be installed. A set of 1500 amp hour batteries is available and will be installed along with the rest of the AC/DC electrical system. After that, research will be directed towards providing electricity with solar thermal steam generator with the focus of a ring of concentrators being at the top of the monolithic chimney. The steam generator will be able to receive its push from any of several available sources including the sunlight concentrators, a giant liquid candle with three inch diameter ropes floating and burning in a giant vat of oil or alcohol, or automatic stoking of wood or coal.

Winterstash Food Coop/Community Greenhouse

The Winterstash Food Cooperative began 18 months ago with a full moon gathering in the park with a handful of people placing a cooperative food order for the winter. Today Winterstash has three hundred members, one-quarter of the population of Telluride and does one-fifth of the gross sales of the only other food market in town. With the little coop bursting at its seams, a new location has been found which will increase Winterstash's available floor space from 200 sq ft to 3,600 sq ft. It is a fine old stone building which will be retrofitted with a solar hot air system. Once again the storage to space heat distribution is strictly by natural convection. Not only will the present retail grocery operations be expanded, but rebagging, rebottling, recycling and community composting will have room for expansion. Helianthus Herbs will share space and a kitchen will be available for restaurant/juice bar service, cooperative drying, canning and bottling operations, as well as providing gathering room space for meetings and classes. There happens to be a big hole in the ground in front of this building which at some later time might be covered to create a coop-greenhouse/community amphitheater. This 48 foot by 72 foot growhole might be covered with a free spanning space frame with pyramid acrylic glazing, an active solar water space heating

Solar Architecture   199

system or perhaps a totally passive heat storage configuration. Insulating curtains will be integral to the space frame. Winterstash deals only in good food, buys from local farmers and mills, and is a member of the Southwest Federation of Co-ops. Winterstash does not support agribusiness policies and is working towards providing the best available food at wholesale prices to the community.

BIOGRAPHICAL SKETCH
Dean Randle is an architectural designer with the Telluride Designworks of Telluride, Colorado.

THE COMMUNITY COLLEGE OF DENVER AND OTHER SOLAR PROJECTS

Presented and prepared by: John Anderson

After analyzing solar energy applications for four years, it is apparent that there is an appropriate technology for each type of architectural project and progress must be made in demonstrating solar energy's applications through the broadest range of project sizes, types and levels of sophistication. The following three projects are currently nearing completion and occupancy. All are relatively large, which makes a strong case for the effective application of solar-assisted heat pump technology.

Community College of Denver/North Campus
The first and largest project is the new North Campus for the Community College of Denver in Westminster, Colorado on the north edge of the metropolitan Denver area. This is the third of the College's three campuses to move into permanent facilities and occupancy is expected in the fall, 1977. Some general statistics are:

Student Capacity	3,517 FTE/Day
Phase I Area	304,950 Gross Sq. Ft.
Building Cost (including solar)	$11,051,000 ($36.24/gsf)
Site Development Cost	698,000
Student Parking Facilities	460,000

During the architectural program development in 1973, an awareness was growing that an energy crisis was either imminent or already upon the metropolitan area. During the winter of 1972/1973, Denver experienced a severe shortage of natural gas and heating oil. It followed with a critically gasoline-short summer, six months ahead of the more widespread shortages of the following winter.

The reality of the energy situation was then brought even closer to home when the Public Service Company of Colorado issued a bulletin to all architects in the state in November 1973 stating that natural gas connections could be assured to new customers only if two very stringent conditions were met. First, applications must be made by 31 December 1973; and second, connections must be made to buildings actually ready for the gas by 30 June 1974. If either of these conditions could not be met, there would be no assurance of natural gas and potential customers would simply have to wait until such assurance might be offered in the future, dependent upon new sources of supply. This reality sparked an immediate study of reasonable alternative sources of energy and within a short time it became apparent that the sun offered the greatest long-term promise as a viable alternative to natural gas.

Four major factors led to the ultimate decision to design the building with solar heating. First was the environmental concern of a beautiful site and the design team's responsibility to it. It was felt that every effort should be made to disturb the site as little as possible and to use the building as a model of non-pollution.

Second was the growing awareness of the finite nature of the natural gas supply. Initial studies showed that as a nation we were entering a non-reversible period of increasing deficit. We were beginning to use more natural gas than could be produced in the future even if such potential resources as liquefied natural gas importation, pipeline importation from Canada and Alaska, and large scale gasificaiton of coal were exploited.

Solar Architecture

Third was the reality that after many years of tight control on the cost of natural gas, the price was about to go up significantly. The most conservative estimate available indicated that the nation was in for at least a 300 percent increase in the cost of gas during the seventeen-year period between 1973 and 1990. Most recent predictions are that a 500 to 800 percent rise is more likely.

The fourth factor favoring the use of solar energy for heating is the fortunate combination of conditions that gives Denver an abundant supply of this resource when it is most needed. Denver has a generally sunny climate year-round and this, coupled with its high altitude and low humidity, gives it more insolation in January than almost any other place in the United States.

Because the College is a state funded facility and it was known from the beginning that initial additional funds would be necessary to design and install the solar heating system, the design team requested that Governor Vanderhoof authorize Bridgers and Paxton, mechanical engineers, to make a study of the technical and financial feasibility of making such an investment. The study was authorized in October, completed in December 1973 and immediately reviewed by the Governor's Science Advisory Committee. The Committee confirmed the findings of the study, endorsed the concept, and recommended that design of the building proceed incorporating a solar heating/heat pump system. Subsequently, the Joint Budget Committee of the General Assembly heard testimony relative to the Governor's request for additional funds, endorsed the concept and included the funding in the capital construction bill of 1974.

The additional funding provided by the legislature for the design and installation of the solar heating/heat pump system amounts to a net increase of 8.47 percent over the funding originally considered for a conventionally designed structure. The major additional cost factors in this increase are: greatly improved insulation, double glazing of all window openings, the solar collectors and their supporting structure, storage tanks for the solar heated water, heat recovery devices on exhaust equipment, and additional controls and operational monitoring instrumentation. Against these additional costs are savings for omission of conventional gas fired boilers and a general decrease in the size of the heating and cooling equipment because of reduced heat loss and gain made possible by better insulation.

The justification for making the additional initial investment is the assurance that there will be substantial reduction in operating cost and that within a reasonable time the investment will be returned to the state through such savings. This "life-cycle cost" concept for many years was not a serious consideration in the funding of new structures because there was an apparent abundance of very cheap fossil fuel and the payback of the cost of going to such alternatives as solar heating could not be proved within the projected useful life of buildings.

However, Bridgers and Paxton's feasibility study showed that based upon the admittedly conservative projection that natural gas cost will increase by 300 percent during the period between 1973 and 1990, the net additional investment for the solar heating/heat pump system will be returned to the state in less than eleven years. Beyond this point, of course, the state will continue to enjoy operating savings each succeeding year. In the year 1990 alone, based on the 300 percent assumption, there could be a savings of over $90,000.

Another of the significant readouts of the feasibility study was the projection of fossil fuel depletion that will occur in the operation of the solar-heated building when compared to the operation of the building that probably would have been designed as little as two years earlier against the same program. In an average winter, the study showed that if the building had been conventionally designed and heated it would consume the equivalent of 74.3 billion Btu's of fossil fuel, gas burned on

Solar Architecture

the site and coal burned at the power plant to generate electricity for pump and fan operation. On the other hand, equivalent fossil fuel consumption for the solar-heated building is predicted at 15.9 billion Btu's per year, all of it coal-burned for power generation (a net fossil fuel depletion of twenty-one percent of the amount for the conventionally designed building).

As designed, the building takes advantage of the natural slope of the site. The upper level is approached from the south with the heavy floor-load industrial shops placed directly on grade and accessible from the main drive system. The site then drops away sharply to the north permitting a second level to develop naturally. The grade is also somewhat lower at the west end so that a third and lowest level fits comfortably into the site. In spite of its 1000 foot plus length, the building is very compact and takes maximum advantage of the potential for tucking the lower levels into the hill, thereby minimizing the area of exposed exterior envelope.

Although the south bank of collectors is supported on a free-standing support system, the north bank covers the south face of a series of eight fan rooms and high-bay skylighted areas over stairways, open wells and public spaces. These spaces give a sense of openness and a welcome change of scale in an otherwise very tightly and efficiently designed structure.

In simplistic terms, the heating system works like this: The sun's energy is captured in a series of solar collector panels totalling some 35,300 square feet mounted on the roof in two banks and facing due south at a tilt angle of approximately 53 degrees to the horizontal. Water is pumped through the panels and the energy is transferred to the water in the form of heat. The heated water is stored in underground tanks having a capacity of approximately 200,000 gallons.

When the temperature of the stored water exceeds 100 F, it can be pumped directly to the heating coils of air-handling units where heat is transferred to air and the air is distributed by fans in conventional fashion through a ductwork system to the various spaces in the building. When the temperature of the stored water is between 55 and 100 F, it is pumped through the chiller-heat pump units where its heat is extracted, consolidated, and sent out in 100 F water to the air-handling units. There is no back-up system as such, but if there is an extended period of hostile weather and the stored water temperature falls below 60 F the gas-fired domestic hot water boilers can be used between 11 pm and 7 am to put an additional day's stored energy into the tank.

With the sole exception of the collector panels, all of the components of the heating system are of standard manufacture and proven performance. The chiller-heat pumps are conventional package-type centrifugal chillers. Selection of the collector contractor was made by analyzing relative solar simulator data from performance tests run by the Lewis Research Laboratories of NASA in Cleveland, comparing the design proposals of ten bidders, and comparing their fixed price bids submitted in October 1975. The collector contractor is Southwestern Sheet Metal Works, Inc. of El Paso, Texas.

In addition to the solar heating system, the building will incorporate a number of additional energy conservation features. During much of the heating season, the heat pumps will act as means of redistributing the generated heat of people and lights within the building, taking it from areas where there is an overabundance and distributing it to areas where it is needed. Heat recovery devices will also be used at the exhaust fans to extract heat that would ordinarily be lost with the exhausted air and return it to the heating system. During the summer, spring and fall -- and much of the winter -- the solar collection system will also provide energy for the heating of domestic water.

Other Solar Projects
Two other projects being completed in the fall, 1977 are virtually identical elementary schools in Aurora, Colorado. Each

one is approximately 49,000 gross square feet in extent and has 2,000 square feet of flat-plate solar collectors. Space behind the two collector banks is utilized for mechancial equipment. Each building is heated by a solar-assisted heat pump system with storage provided by two underground tanks of approximately 14,000 gallon total capacity. Electric boilers provide back-up as necessary.

The schools are designed on an open-plan flexible-space concept with demountable partitioning that can be rearranged over the years to respond to changes in the educational program. Central to the academic area in each school is a high-bay instructional materials center with brightly painted heating pipes and ducts. As in the case of the Community College, as much as possible of the heating system is exposed and understandable.

Three additional solar-heated projects in Colorado are currently under design by the firm. One is a 2,500 square foot residence with an air collector system and pebble-bed storage. A second is a small church in Pueblo utilizing a relatively small collector area and a larger-than-average storage tank to respond to the intermittent use pattern of the facility. The third project is a dormitory facility for 245 single students at the Colorado School of Mines in Golden with twin five-story towers linked by a steel space frame truss at roof level that supports the solar collector array.

Each new project calls for a fresh approach to solar application. There is much yet to be learned and to be explored, but with each new attempt, new answers are worked out and new and exciting concepts emerge. The firm, John Anderson and Associates, is committed to continue the work and is enthused about the possibilities.

BIOGRAPHICAL SKETCH
John Anderson is a well-known architect from Denver, Colorado and is president of John Anderson and Associates. Mr. Anderson is and has been very active professionally and politically with the advancement of energy-conserving and solar-oriented architecture.

BASELINE DESIGN OF COMMERCIAL CENTRAL RECEIVER SOLAR POWER PLANT

Presented and prepared by: Floyd A. Blake

INTRODUCTION

Parallel programs to establish the technology base and optimum performance-economic approach for utility scale solar electric power generation have been underway under National Science Foundation and Energy Research and Development Administration sponsorship since 1973. Martin Marietta together with teammate organizations, Bechtel Corporation, Foster-Wheeler Energy Corporation and Georgia Institute of Technology has concentrated on the central receiver configuration rather than a distributed energy collection configuration due to its higher performance and economic potential.

The initial NSF sponsored "Solar Power System and Component Research Program" (Grant AER 74-07570) resulted in a conceptual point design 100 MW_e Solar Power Plant[1] and preliminary design of a 1 MW_{th} solar steam generator which included the basic features of the full scale unit and adaptation features to enable operation in the Centre National de la Recherche Scientifique solar furnace at Odeillo, France[2]. Based on the results of this and parallel programs, the Energy Research and Development Administration established central receiver solar power generation as one of its major R and D activities. In June 1975 three Phase I system contracts and one Phase I subsystem contract were initiated. The system design contracts were to run two years and "Establish the technical feasibility of a solar thermal power plant of the Central Receiver type with significant commercial potential" and obtain sufficient development, production and operating data to indicate economics of operation for commercial power plants of similar design.

Specific segments of the Phase I "Central Receiver Solar Thermal Power System" program (ERDA Contract EY-76-C-03-1110) are:

1. Conceptual baseline design of commercial plant, 10 MW_e Pilot Plant, and Solar Subsystem Research Experiments.

2. Detail design, fabrication, and test operation of solar subsystem research experiments.

3. Preliminary design of the 10 MW_e Central Receiver Solar Thermal Power Pilot Plant.

4. Costing of the 10 MW_e pilot plant and large scale commercial solar power plant.

This paper summarizes the results obtained during the first 16 months of the Phase I Central Receiver Solar Thermal Power System program being performed by the Martin Marietta, Bechtel, Foster-Wheeler, Georgia Tech team during which the work effort of tasks 1 and 2 was largely completed. Tasks 3 and 4 will be performed in the final months of Phase I.

SOLAR PLANT AND EXPERIMENT DESIGNS

Commercial CRSTP Plant
Two conceptual configurations of the commercial size central receiver solar thermal power plant have been established during the program. The first, generated at the outset of the program was rated at 100 MW_e solar and 70 MW_e storage, with the storage capacity being six hours. The second configuration established during the initial stage of the preliminary design effort

Solar Architecture 211

reflected the increased knowledge base generated during the first year of the program and revised guidelines established by the customer. The updated commercial plant configuration is rated at 150 MW$_e$ solar and 105 MW$_e$ storage with the storage capacity being three hours. Figure 1 shows a comparison of the two plant layouts and key parameters.

Basic features of the Martin Marietta Team commercial solar power plants have been established to achieve the highest performance consistent with minimized capital and operating costs, and timely development of solar power technology. Strongly contributing to these goals is the modular collector-receiver which provides maximized optical performance and thermal energy conversion efficiency. Within the modules, the north field collector of focused heliostats transmits focused sunlight over moderate slant ranges maximizing optical performance. The cavity receiver is designed for maximum performance with flux levels consistent with commercial power system steam generators. The

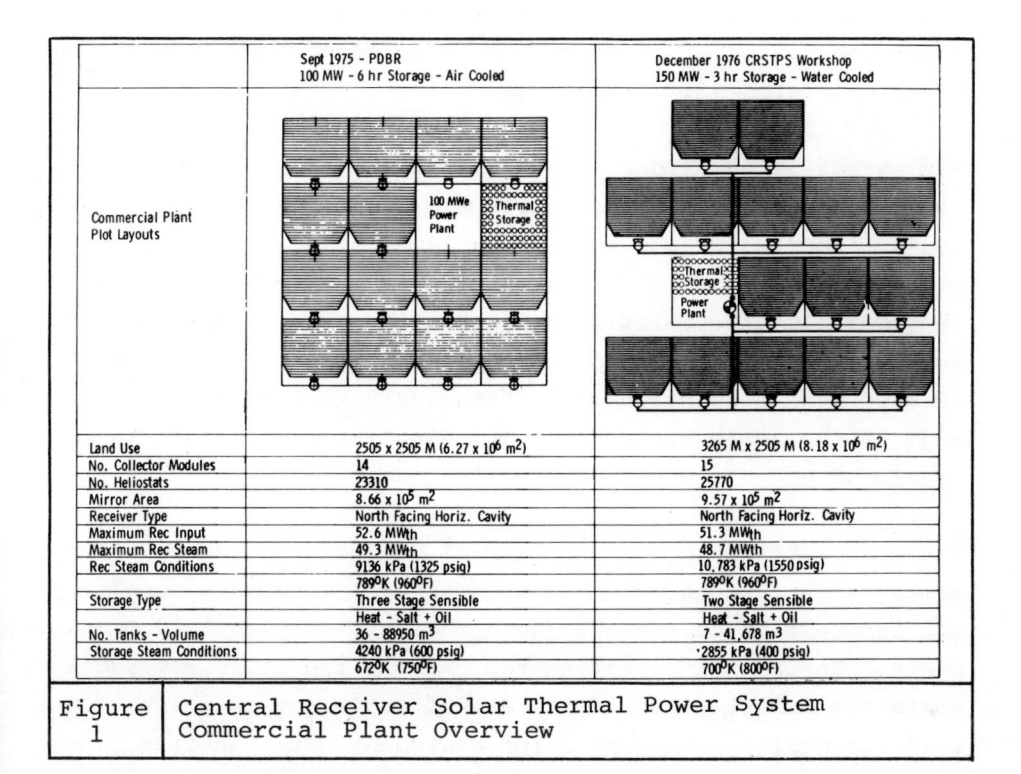

	Sept 1975 - PDBR 100 MW - 6 hr Storage - Air Cooled	December 1976 CRSTPS Workshop 150 MW - 3 hr Storage - Water Cooled
Commercial Plant Plot Layouts		
Land Use	2505 x 2505 M (6.27 x 10^6 m^2)	3265 M x 2505 M (8.18 x 10^6 m^2)
No. Collector Modules	14	15
No. Heliostats	23310	25770
Mirror Area	8.66 x 10^5 m^2	9.57 x 10^5 m^2
Receiver Type	North Facing Horiz. Cavity	North Facing Horiz. Cavity
Maximum Rec Input	52.6 MW$_{th}$	51.3 MW$_{th}$
Maximum Rec Steam	49.3 MW$_{th}$	48.7 MW$_{th}$
Rec Steam Conditions	9136 kPa (1325 psig) 789°K (960°F)	10,783 kPa (1550 psig) 789°K (960°F)
Storage Type	Three Stage Sensible Heat - Salt + Oil	Two Stage Sensible Heat - Salt + Oil
No. Tanks - Volume	36 - 88950 m^3	7 - 41,678 m^3
Storage Steam Conditions	4240 kPa (600 psig) 672°K (750°F)	2855 kPa (400 psig) 700°K (800°F)

Figure 1 Central Receiver Solar Thermal Power System Commercial Plant Overview

shortened tower is the one clear case where economics benefitted at the expense of energy collection efficiency, with the savings in tower costs being greater than the cost of extra heliostats required to replace the reduced capacity.

Modular design of the collector-receiver module provides versatility of design in that site variations and sizing variations over broad limits can be accommodated by the basic design. The modular design also provides the shortest development path to the commercial plant in that all solar subsystems are full scale by the pilot plant stage. Key also to attaining early development and commercial acceptance are the low risk features of the design, notably the cavity receiver steam generator, the focused heliostats, use of sensible heat storage and use of a single-stage expansion turbine. The updated commercial plant design has an end to end efficiency of 25 percent while operating on receiver steam, based on the power delivered to the transmission line divided by the potential solar energy intercepted by the full area of the mirrors in the system. This is a specific power generation of 238 W/m^2 (for reference a spacecraft solar array generated 93.2 W/m^2 in equivalent terrestrial sunlight).

Key guideline revisions included changing from air cooling to water cooling, a decrease in the storage time from 6 to 3 hours and an increase in the insolation level at design point from 0.825 to 0.95 KW/m^2. During the design update and optimization the combined effect influenced selection of a commercial plant configuration with a 50 percent higher rating which requires 10.6 percent more heliostats (from 23310 to 25770). The arrangement of the collector modules and plant-storage module is a result of piping loss/cost optimization.

Factors contributing to the overall economics of scale of the commercial plant include elements of cost which are sensitive and insensitive to scale in both capital equipment and operating costs. In capital equipment for example the collector-receiver costs are direct linear functions of scale (or constant $/KW) while electric power generation equipment cost benefits from

Solar Architecture

increased scale and piping costs suffer from increased scale. Operating costs associated with plant operation benefit from scale while maintenance costs of the collector field do not.

The 150 MW_e plant sizing is based on a trade-off of the scale sensitive factors. Conversion of operating costs to equivalent capital costs was made to enable comparison on the same baseline, dollars per kilowatt capital cost. The combined effect of the major scaling influences, the turbine generator's declining cost with scale, the piping's rising cost with scale, and the operating manpower's declining cost with scale is a very flat curve in the 100-150 MW_e range followed by a break upward above 150 MW_e. 150 MW_e was selected for the commercial plant size on the basis of the foregoing combined influences study: the recognition that fixed plant facilities, not yet considered, will also benefit from larger scale, and that a 150 MW_e turbine generator can be assembled from existing hardware.

Figure 2. Pilot Plant Artist Concept

10 MW$_e$ PILOT PLANT

The pilot plant, shown in the artist concept of Figure 2, is basically a collector-receiver module of the commercial plant integrated with suitably sized thermal storage and electric power generation subsystems. This is the key to timely development and commercialization of the central receiver solar power technology since it virtually eliminates any scaling difference between the commercial and pilot plants in the solar subsystems. Timely development is further enhanced by use of a conventional moderate pressure single expansion turbine and use of commercial heat transport fluids in the thermal storage subsystem.

The collector subsystem for the pilot plant/commercial module contain 1718 heliostats and is focused into the cavity receiver mounted on the tower with the aperture centerline at 90 meters. This provides one-for-one scaling with the commercial plant including all of the performance characteristics of the collector subsystem as well as all receiver subsystem performance characteristics. This includes the heat flux and distributions and total heat into the cavity throughout the the year, steam temperature, pressures and flowrates for all operating modes using the controls hardware and configuration which is the same as the commercial plant. This size collector provides a pilot plant operation at 10 MW$_e$ net while simultaneously charging storage at a moderate rate, simulating commercial plant operation.

The thermal storage subsystem is functionally and operationally the same as the commercial plant and maintains the important scaling criteria including heat transfer rates, storage media fluid temperature, heat exchanger coefficients, and steam temperatures and pressures into the turbine admission point.

The electrical power generation subsystem used for the pilot plant includes an admission turbine, as does the commercial plant, and has all of the operational and control modes of the commercial plant. The pilot plant turbine size has been scaled down from the utility size turbine in the commercial plant to an

Solar Architecture   215

industrial size.

COLLECTOR SUBSYSTEM RESEARCH

A collector research experiment has been designed and built under the Phase I program to initiate the development of the vital solar energy collection/concentration component, the focusing heliostat. Shown in Figures 3 and 4 are the two configurations designed for the experiment, a 25 facet "test heliostat" with variable focusing capability and a 9 facet "commercial concept heliostat" with pre-focused mirrors. Common features of both

Figure 3 | 25 Mirror Heliostat at SRE Position 11

Figure 4 | 9 Mirror Heliostat Prefocused for 400 M

Figure 5 | 4 Heliostats in Operation from L to R

Figure 6 | Flares from 4 Operating Heliostats

| Figure 7 | Calorimeter-Doors Open |

| Figure 8 | Calorimeter-Doors Closed |

| Figure 9 | Commercial/Pilot Plant Receiver Plan Section |

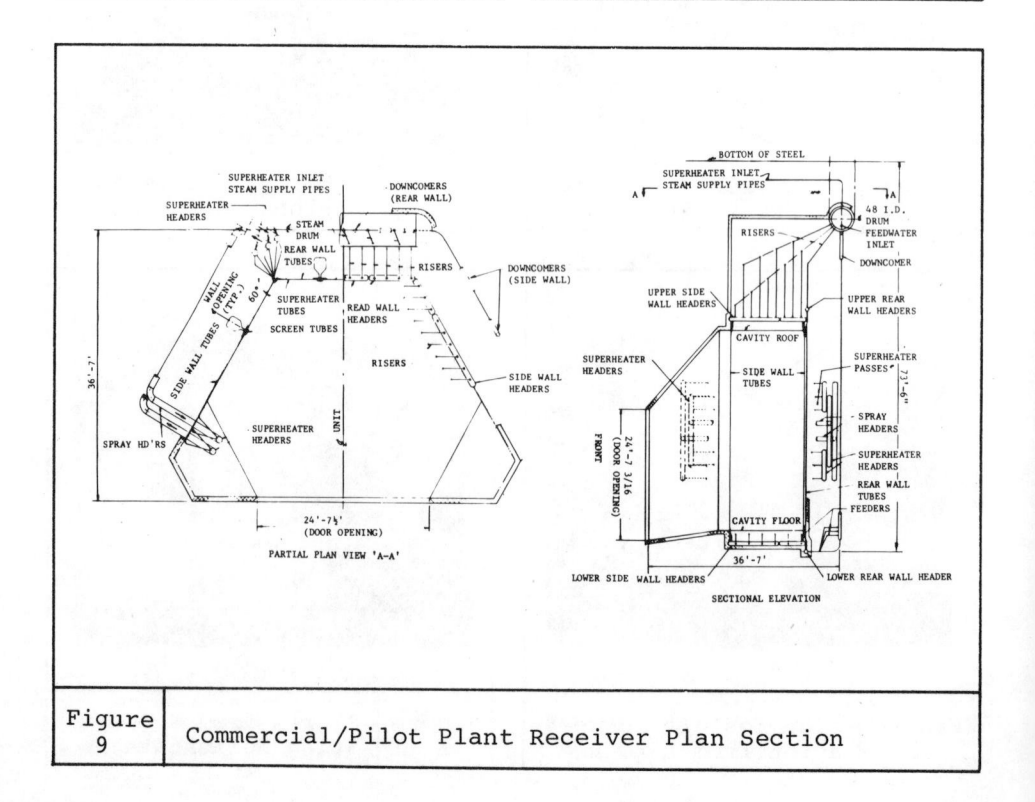

Solar Architecture

heliostat designs are azimuth-elevation tracking mechanism, 39.2 m^2 (400 ft^2) mirror area, face down storage and closed loop tracking control based on pointing error signals from an in-line sensor.

Three units of the 25 facet variable focus design and one unit of the 9 facet pre-focused design have been installed in a valley on the Martin Marietta plant site as shown in Figures 5 and 6. A ridge across the southern end of the valley provides the tower height for the test site and is the location for the energy measurement equipment shown in Figures 7 and 8.

Early test results from the heliostat subsystem research experiment have verified the gaussian energy distribution across the focused energy from a concentrating heliostat, cross checked radiometric and calorimetric measurement techniques within 3 percent and established the atmospheric loss between the heliostat and receiver to be approximately 3 percent at 342 m (1123 ft) slant range.

RECEIVER SUBSYSTEM RESEARCH EXPERIMENT

Plan and elevation drawings of the cavity receiver steam generator for the commercial and pilot plants are shown in Figure 9. The cavity concept has an inherent geometric advantage in absorbing a high percentage of incident radiant energy. Radiation and convective heat losses are minimized by the enclosure surfaces. Furthermore, overnight shutdown losses are minimized by implementing a remote operated door to cover the aperture.

The natural circulation boiler concept has demonstrated a long history of highly reliable operation in steam power plant applications. The characteristics of a natural circulation unit to be self-compensating for variations in energy input are particularly advantageous in the solar plant application, wherein we can expect nearly continuous (and at times abrupt) changes in heat flux patterns on the absorbing surfaces. The resulting simplicity in flow circuitry, valving and controls is most conducive to minimizing overall risk factors for the Central Receiver Solar Thermal Power Program. Since our design point temperature, pressure and heat flux levels are well within the demonstrated state of the art for natural circulation boilers, our receiver design carries a high confidence of providing long lifetime reliable operation.

The subsystem research experiment consists of a 5 MW_{th} cavity receiver steam generator installed in the required structure to enable testing at the ERDA/Sandia Solar Thermal Test Facility and its support and instrumentation systems. Design, fabrication, and erection of the receiver were accomplished by Foster-Wheeler Energy Corporation.

Testing during the current program will use heat from an IR lamp array of the Sandia Radiant Heat Laboratory. A model of the 5 MW_{th} receiver is shown in Figures 10 and 11 and the unit under construction is shown in Figures 12 and 13.

Principal differences in the configuration between the 5 MW_{th} SRE receiver and the commercial/pilot plant receiver resulted from recent design analyses. The structural support will be from the tower enclosure in the commercial design primarily to reduce the effects of seismic loads at the receiver/tower interface. The tilt requirement for operation at the Solar Thermal Test Facility and in the conceptual baseline design was eliminated as a secondary benefit of the tower-height reduction to 90 m.

The operating pressure level of the receiver was increased from 9136 kPa (1325 psig) to 10783 kPa (1550 psig) for system integration reasons. The change enabled use of a pressure balanced, cost-minimized piping system and slightly improved the heat rate of the turbine. Natural circulation boiling has been preserved, but the design is approaching the limit for this type operation at the flux levels of a high performance collector. Checkout testing of the 5 MW_{th} SRE receiver was initiated in early February 1977.

Figure 10 — Front view showing Superheater panels

Figure 11 — Profile showing Drum, Risers & Downcomers

Figure 12 — Front view, under construction

Figure 13 — Rear view, prior to insulation & enclosure

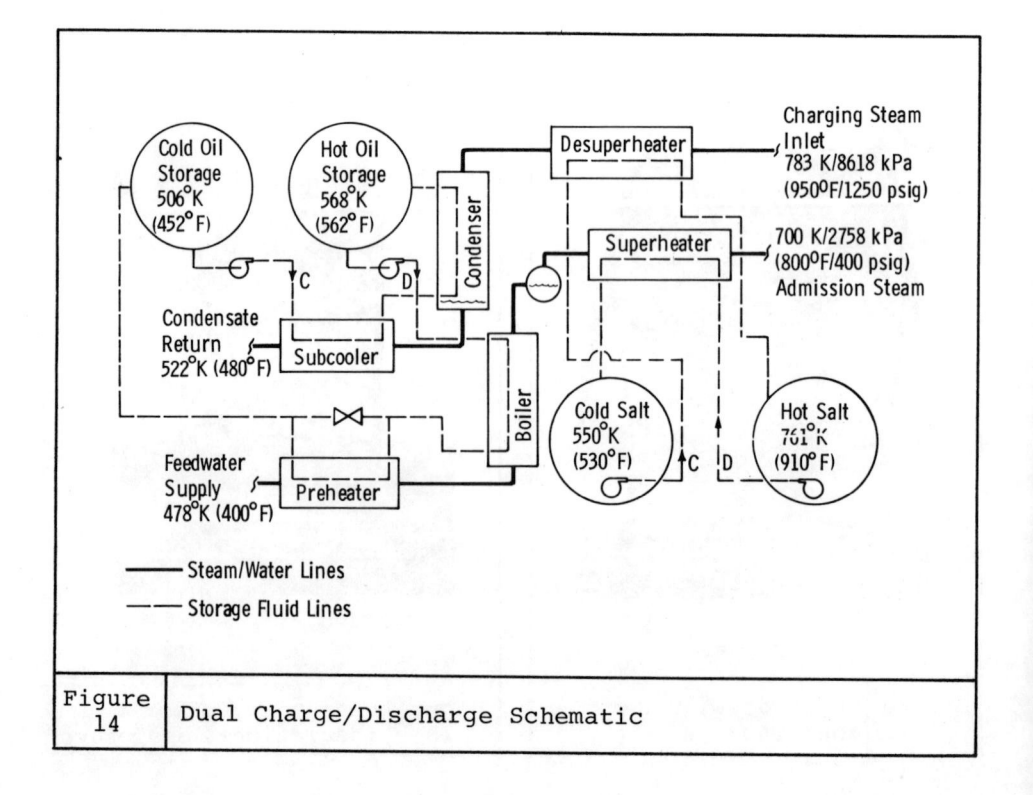

THERMAL STORAGE SUBSYSTEM RESEARCH EXPERIMENT

The thermal storage program philosophy has been to design and demonstrate a complete thermal storage subsystem with a minimum technical and developmental risk. A technically well understood liquid sensible heat storage system was selected. This system allows us to apply existing heat transfer technology and use commercially available components and controls. The TSS, shown schematically in Figure 14, stores "superheat" energy in molten salt at 761°K (910 F) and "vaporization energy" in hydrocarbon oil at 568°K (562 F). During the charge cycle 783°K/8618 kPa (950 F/1250 psig) steam direct from the receiver enters the desuperheater heat exchanger where it heats molten salt while giving up its superheat energy. The steam then enters a condenser where it gives up its heat of vaporization while heating storage system oil. A subcooler heat exchanger preheats oil going to the condenser while dropping the water to a compatible temperature for the receiver feedwater pumps.

Figure 14. Dual Charge/Discharge Schematic

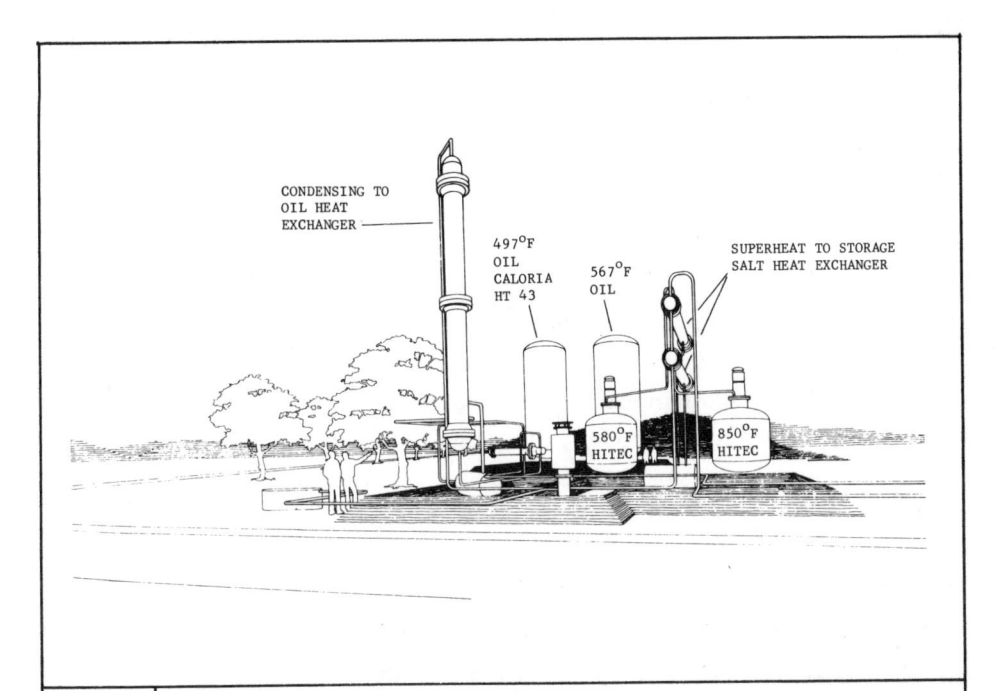

Figure 15. Perspective View of Thermal Storage Subsystem Research Experiment (Without Supporting Structures)

Figure 16. View of Thermal Storage Subsystem Research Experiment Under Construction

The discharge cycle is a conventional preheat, vaporization, superheat sequence performed in heat exchangers. Oil from the hot tank initially vaporizes the feedwater in the boiler and then is used in the preheat section. Molten salt from the hot tank powers the superheater. Parallel charge and discharge capability was incorporated into design when the magnitude of the design constraint placed on heat exchangers by the ½ hour turn-around was established. The acceptable thermal cycling rate for the large heat exchangers of an optimum design precluded attaining the ½ hour turn-around. A change to a longer turn-around time was operationally unacceptable and the dual mode feature with zero turn-around time was therefore selected.

The thermal storage subsystem research experiment was designed and built by Georgia Institute of Technology. The site is the Georgia Power Company, Plant Yates. The thermal storage experiment is shown in the perspective drawing of Figure 15 and in the photographs of Figures 16 and 17. Checkout testing of the unit,

Figure 17 Desuperheating-Superheating Heat Exchanger HE 1

Solar Architecture

rated at 1.6 MW$_{th}$-hr capacity was completed in early February 1977.

ELECTRIC POWER GENERATION SUBSYSTEM

The design of the electric power generation subsystem, tower for the central receiver, and balance of plant has been performed by Bechtel Corporation. The major update modification of the EPGS has been the increase of the commercial plant rating from 100 to 105 MW$_e$ net while operating on receiver steam and from 70 to 105 MW$_e$ net while operating from storage steam. In accordance with design guidelines the method of waste heat rejection was changed from air-cooled to water-cooled condensing. Operating state pressures and temperatures have been modified as needed by system integration considerations while staying within the state of the art of existing hardware.

The depth of design has reached the point where specific power generation equipment has been selected. A cross section of a

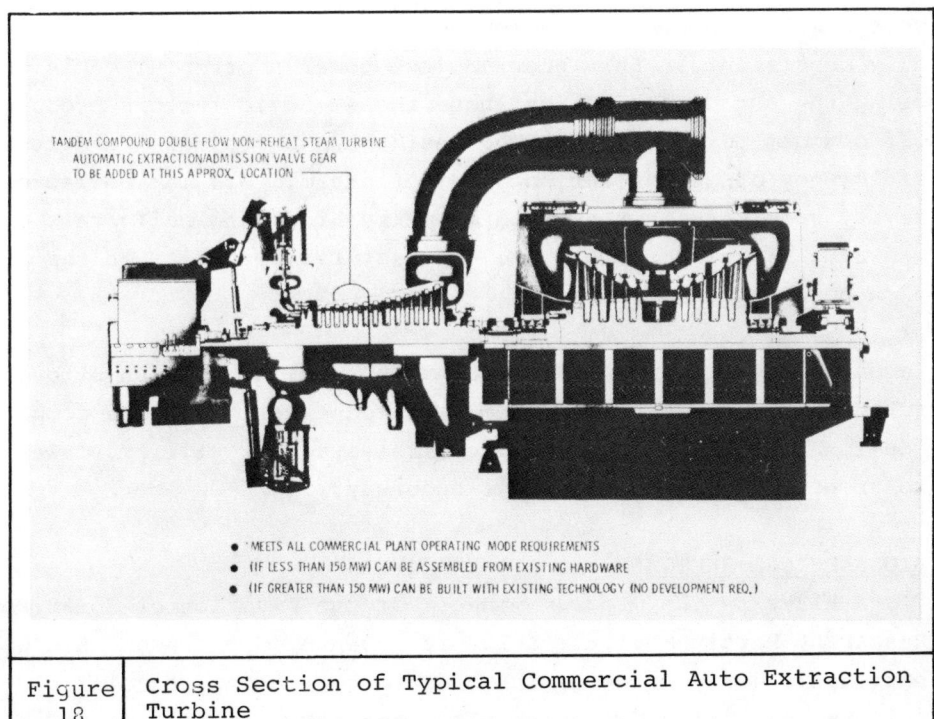

Figure 18. Cross Section of Typical Commercial Auto Extraction Turbine

typical tandem compound double flow non-reheat turbine is shown in Figure 18 with the location of the admission gear for storage steam indicated. The pilot plant electrical power generation subsystem is capable of all of the operational modes of the commercial plant, therefore permitting complete simulation. The required change from dry to wet cooling in both the pilot and commercial plants provided increased efficiency of plant performance.

CONCLUSIONS

At this stage, 18 months into the 24 months Phase I Central Receiver Solar Thermal Power System Program, the accumulated integrated team effort has produced a Commercial Conceptual Design having substantial advantages over competing concepts both within and outside of the Central Receiver concept.

A constant goal and guideline throughout the program has been the optimization of performance within the framework of optimum economics. These two goals have been demonstrated to be consistent in power generation technology and have been generally consistent in this program (the tower shortening decision being the notable exception where the economic factors produce off optimum performance). The design has a projected end-to-end efficiency of 25 percent on receiver steam. Modular collectors, cavity receivers, north field geometry, focusing heliostats, and sensible heat storage are key features resulting in the high-performance potential.

Technical feasibility of the solar subsystems has been strongly substantiated by the subsystem research experiments and companion 1 MW receiver program. The first stage of utility scale solar power generation is well underway.

BIOGRAPHICAL SKETCH

Floyd Blake is the program manager of the Division of Solar Power Research, Martin Marietta Aerospace, Denver, Colorado. His work has mainly dealt with the large scale generation of solar electricity (e.g., 100 MWe solar plant design configuration and

Solar Architecture

performance), cadmium sulfide cells and cavity generator systems.

ENDNOTES
1. F.A. Blake, "Central Receiver Solar Thermal Power," Presented to Winnipeg Solar Energy Conference (U.S. and Canadian Sections of International Solar Energy Society) August 1976.
2. F.A. Blake, T.R. Tracey, J.D. Walton, S.H. Bomar, Jr., "One MW_{th} Bench Model Cavity Receiver Steam Generator," Presented to American Section, International Solar Energy Society, Los Angeles, July 1975.

WIND POWER AS A VIABLE ENERGY SOURCE

Presented by and prepared by: Stan H. Lowy

The use of wind for mechanical power dates back to at least the early 12th century and perhaps as early as the 10th century. Persia seems to be the origin of the first windmills which were used to grind corn. The mill had a horizontal axis with sails extended over radial arms. The idea of utilizing the wind for mechanical power spread, probably by means of the soldiers and prisoners of Genghis Khan, to China, where the most prevalent use was for irrigation. By the 14th century, windmills were in use throughout Europe except for Spain, where they finally came into use about the mid-15th century. Each region utilized its own particular design and made improvements to suit the particular needs of the region.

In the United States, windmills initially were used primarily for pumping water from wells. Companies such as Dempster, Baker and Aermoter sold thousands of units for use throughout the Southern, Midwestern, Southwestern and Western regions of the USA. In the early 1900's, following the development of automotive electrical systems, small DC generators became available and, with minor changes, were used extensively in wind driven systems. A contributing factor here was the research and development of aerodynamically improved propeller blades. The

1920's and 1930's saw large-scale use in rural areas of wind driven electric systems, ranging from about 200 watts to about 3,000 watts. But with the formation of the REA (Rural Electrification Agency), the rural areas were able to obtain large amounts of electricity more cheaply and with more convenience. Hence, wind driven electrical systems almost disappeared. However, approximately 150,000 windmills are still in use in the United States for pumping water.

Emphasis on oil and gas conservation and talk of energy independence of the USA, with regard to foreign oil supplies, about two years ago prompted a considerable number of efforts to utilize other forms of energy, such as coal, nuclear power, geothermal heat and direct energy from the sun for use in the electric power industry. At the present time, the utility industry is heavily dependent upon fuel oil and natural gas as the primary energy supply. About 10% of our oil consumption and 16% of our gas consumption is used to generate electricity. The problems of increasing costs for fuel oil, natural gas, atomic reactor construction and environmental considerations of the uses of high-sulpher content coal and plutonium fuel, are fraught with political, social and economic overtones. For example, Con-Edison's projection for uranium in 1985 was $34 per pound. But in August 1976, the cost was $42 per pound. In addition, their nuclear units, in 1975, ran at only 33.4% of capacity compared to 52.8% of capacity for coal units.

One method of producing electrical power, which simultaneously conserves oil and gas and causes no pollution, has been given very low priority and yet it may be the most appropriate method of electrical power production. That method is the use of wind energy to rotate a driving device known as a turbine, such as a propeller, which in turn drives a generator. All the technology necessary for the design and construction of such a device has been available for many years. The fuel necessary to drive the device, the wind, is free. The driver, generator and other required equipment cost is little more than conventional generating plant equipment.

Solar Architecture

WIND TURBINES

Wind turbines may be classified as horizontal axis or vertical axis types. Examples of horizontal axis turbines are the propeller blade turbines having one or more propeller blades fastened to a hub, such as the familiar farm multibladed windmill, and the sail wing. Examples of vertical axis turbines are the Savonius blades, the Darrieus blades and the turbine. The Savonius type is primarily a drag device and is self-starting with low rotational speed and low efficiency. The low tip speed ratio results in low power output per unit cost. The Darrieus type is primarily a lift device which is not self-starting but attains high rotational speeds, once started, and has a high efficiency. It has a potential for low capital cost but a larger diameter is required per power output than for the propeller type.

The use of wind-driven generators can reduce the consumption of fuel oil and natural gas by taking on the burden of power production except for those days when the wind is of insufficient energy or of excessive energy, i.e., in excess of about 60 to 80 miles per hour. If an energy storage system has been designed into the wind-driven generating plant, then power can be generated even on windless days without resort to conventional fuels.

Nineteen-seventy-five energy consumption in the USA was about 2.5×10^{12} KWH per year (2.5 trillion KWH per year). Of this, about 0.625×10^{12} KWH per year (or 25%) was consumed to produce electricity. Another 25% was used for industrial processing, 24% for transportation, 19% for heating and air conditioning buildings, 5% for petrochemicals and 2% for other products. Energy sources are oil (40%), natural gas (33%), coal (20%), hydropower (4%), nuclear (2%) and other forms (1%). In 1974, the USA imported 6×10^{6} barrels of oil per day and approximately 14% of our energy consumption came from that imported oil. In 1976 this increased to about 50%. Hence, to replace this amount with wind generated energy would have required units

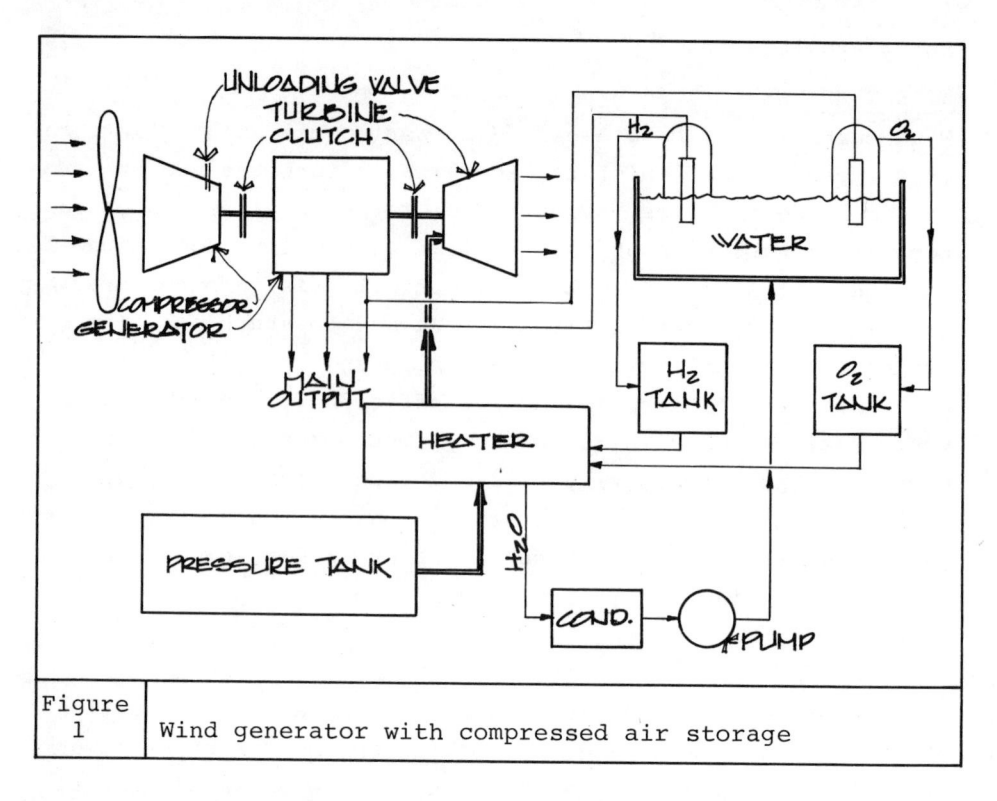

Figure 1 — Wind generator with compressed air storage

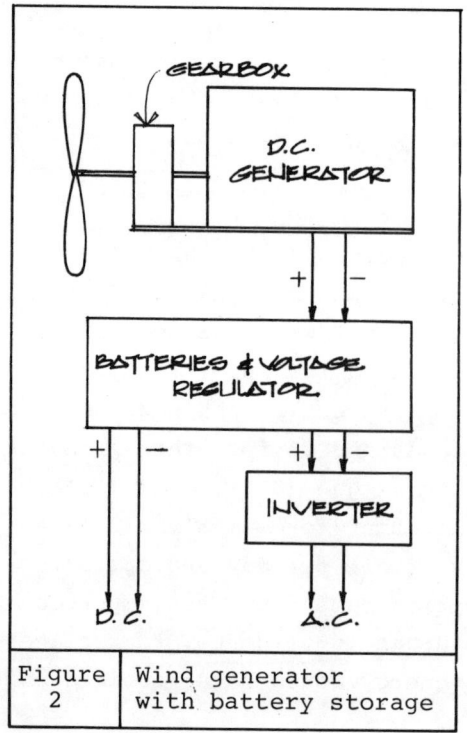

Figure 2 — Wind generator with battery storage

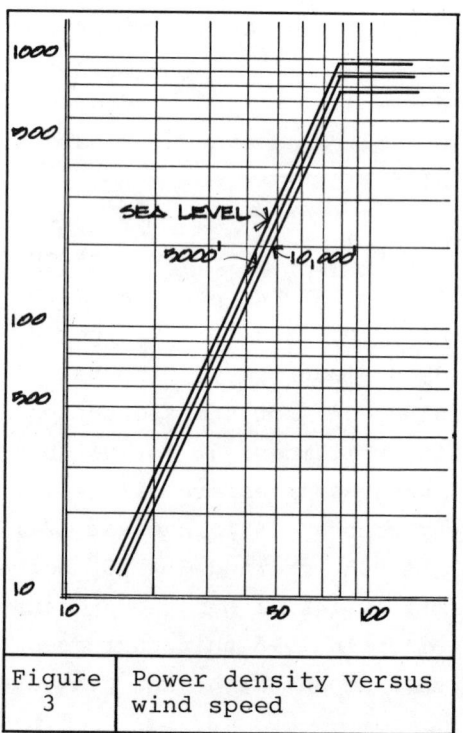

Figure 3 — Power density versus wind speed

Solar Architecture

with a total capacity of 3.5×10^{11} KWH per year in 1975 and about 13×10^{11} KWH per year in 1976. This was approximately 3.7 times the total US energy consumption in 1900. By the year 2000, the US energy consumption is expected to be the equivalent of about 6×10^{12} KWH per year.

For 1970, electricity use for farm production and family living amounted to 37×10^{9} KWH per year. For rural residential heating in 1970, electricity represented 13.8% of the energy expended and for farm production and family living, electricity represents 8.53% of the total energy expended.

In 1972, the Solar Energy Panel of the NSF and NASA estimated that the power potentially available from surface winds over continental USA, the Aleutian Arc and the Eastern Seaboard is equivalent to approximately 10^{5} gigawatts of electricity. This is over 30 times more than the estimated total power consumption in the US by 1980 and more than 100 times the estimated electrical generating capacity in the US by that time.

Ultimately, it will be necessary to fully utilize the inexhaustible, although intermittent, non-polluting sources of power such as sunlight, winds and tides. Renewable energy resources permit a level of energy independence for any country. The problem is one of storing this energy in a form as convenient as liquid or gaseous fuel or even coal. At the present time storage of excess wind energy is being suggested in the following forms:

- Batteries (approximately $50/KWH capacity)
- H_2 and O_2 (from electrolysis of water)(approximately $800/#H2/hr capacity)
- Water storage (as hot water in tanks or pumped back behind dams)
- Fuel cells (approximately $75/KWH capacity)
- Compressed gas (approximately $150/KWH capacity)

Additional uses for wind-driven turbines are to pump water from wells; to reduce wind velocity behind the rotor by absorbing wind power; producing hot air, by driving a compressor, for uses such as building heating and crop drying; as an energy source for fertilizer manufacturing, direct reductions of iron ore and in aluminum production; and as an energy source for pumping irrigation water from ponds or wells.

Wind Power Design

The maximum power available from the wind is $16/27 \cdot 1/2 \, \rho AV^3$ or $0.2963 \, \rho AV^3$ where ρ is the air density in slugs per cubic foot (1 slug/ft^3 = 32.17#/ft^3), A is the swept area of the turbine in square feet (= $\pi/4$ times the square of the diameter) and V is the wind speed in feet per second.

Air density does not change appreciably at any particular site; at most the variation is less than 1.5%. The swept area can be changed only by physically changing the turbine, e.g., replace a 50 foot diameter by a 60 foot diameter blade. But the wind speed varies appreciably from day to day, even hour to hour, and the power obtained is proportional to the cube of the velocity, e.g., doubling wind velocity results in an increase of power by a factor of eight (Figure 3).

Wind generators are designed on the basis of average annual wind speed and this does not vary at a given site by more than about 20%. Generators should be located in areas of the safest, steadiest and fastest wind speeds, and so careful site selection is necessary. In addition, winds are generally stronger at 100 to 200 feet elevations than at ground level, thus suggesting the use of towers to elevate the turbine into higher energy winds.

The choice of a site for the windmill is of utmost importance. The average wind speeds during a year's time must be determined in order to calculate the output (KWH/month) of a particular size wind-driven generator. This information is available from the Weather Bureau, and although they may not have information for the exact site, they probably have records for a nearby region.

Solar Architecture

It is also quite helpful to know the longest period of no wind and the shortest period of some wind. An adequate site for a wind-driven generator exists if there is a 10 mph average wind over a period of two or three days per week. For example, a 3 KWH generator, at 115 volts, will produce 400 to 500 KWH per month when driven by a 15 foot diameter propeller at 225 rpm in 10 to 20 mph winds for two or three days per week.

Power can be produced at wind speeds as low as 8 or 10 miles per hour when propeller blades are at their lowest pitch setting. As wind speed increases, the angular velocity (revolutions per minute) can be made to remain constant by using a comparatively simple and long-lasting governor system which increases the blade pitch and thus allows the increased torque applied to be absorbed. A disadvantage of this system is that the power coefficient is reduced because of a reduction in the tip-speed ratio. At about 60 miles per hour, the pitch angle will have maximized and any further increase in pitch will cause blade stall and a sudden decrease in power available.

There are, however, some very preliminary data which indicate that the use of a supercritical airfoil section will delay blade stall to a much higher pitch angle, thus permitting increases in power available until wind speeds are reached in the vicinity of 80 miles per hour. Hence, operation even during stormy periods is a strong possibility. Stormy periods usually will show increased electrical power consumption in a given region compared to non-stormy periods and consequently continuous, and even an increased level of, power production is desirable.

Conservative estimates of the electrical power output (KWH per year per installed KW) possible at eight different Texas locations are shown below. The basis of the estimates are: Cut-in wind speed of 13 mph, rated wind speed 25 mph, shut-down wind speed of 60 mph, and overall power coefficient of .40.

	Location	Specific output, KWH/yr/KW
1.	Amarillo	5,000
2.	Lubbock	4,800

	Location	Specific output, KWH/yr/KW
3.	Corpus Christi	4,500
4.	Brownsville	4,350
5.	Waco	4,300
6.	Wichita Falls	4,250
7.	Dallas-Fort Worth	4,050
8.	Galveston	3,950

A 1000 KW generator installed in the Dallas-Fort Worth area would then produce approximately 4×10^6 KWH per year, or enough to satisfy the requirements of about 500 homes. A similar installation in the Amarillo area could probably take care of 615 homes. About 300 wind-driven generators of 3,000 KW capacity should be sufficient to satisfy the residential electrical requirements of the Dallas-Fort Worth area. These could be installed on various "farms" surrounding the metropolitan area.

The Aspen area potential is approximately 1/3 of that of Amarillo, or about 1700 KWH per year per installed KW. Hence, a 1000 KW machine should take care of 140 homes using about 1000 KWH per month, average consumption. About 18 to 20 1000 KW machines (or 6 or 7 of the 3 MW size) would probably be adequate for residential needs of the city.

On an individual basis, it is readily possible to select a generator and propeller size for electrical energy supply to a home or farm. One starts with a table or chart (graph) of the annual wind speed distribution for the area in which the wind generator is to be erected. This information is available from the Weather Bureau. From this, and the graph of wind machine performance, a graph is constructed of system efficiency as a function of wind speed. The system efficiency is then used to determine the annual maximum power output for various values of generator specific power. The maximum annual output, maximum efficiency, wind speed and rated wind speed are then plotted against specific power in order to determine the annual output and required blade diameter.

| Figure 4 | Annual output and required blade diameter | Figure 5 | Power Schedule |

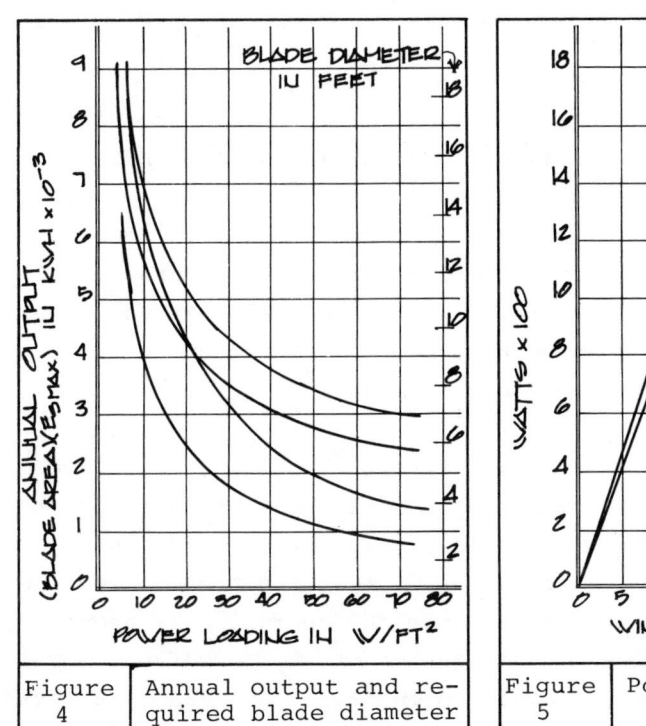

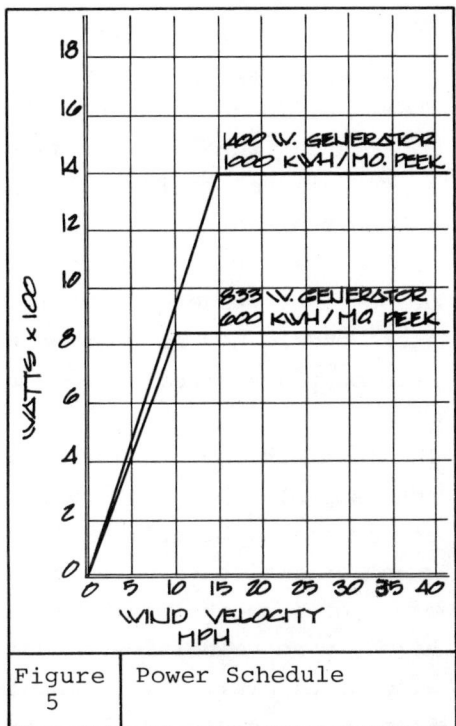

CASE STUDIES

In case A, it is assumed that a peak of 1000 KWH per month is desired with an annual requirement of 7200 KWH per year (average of 600 KWH per month). In case B, a peak of 600 KWH per month is desired with an average of 360 KWH per month, or 4320 KWH per year. In both cases, rated wind speed occurs at 15 mph and the wind speed for maximum efficiency is 13 mph (Figures 4 and 5).

Case A shows a propeller diameter requirement of 14'-6", whereas case B shows a requirement of 11'-4", which is 78% of case A. However, the power requirement of case B is only 60% of that of A, which results from the power being a function of the square of the diameter. In these cases, the specific generator output is 8.4 watts per square foot of propeller area. The resulting power schedule indicates a cut-in wind speed of 7 mph, possibly as low as 5 mph. In case A, a wind speed of slightly less than 11 mph will produce 1000 watts, or 1 KW. In case B, the same wind speed produces about 600 watts.

OTHER CONSIDERATIONS

Tower design and installation is extremely important and it is recommended that the design be checked, if not accomplished, by a professional engineer. Wind generators should <u>not</u> be mounted on the roofs of homes because the loads on the turbines may easily reach values which can cause serious structural damage or outright structural failure. Even small wind-driven generators should not be roof-mounted because of the noise and vibration transmitted through the structure.

The topography of the site will affect the wind velocity to a considerable extent (Figure 6). For example, trees in front of, or behind, a mill will seriously interfere with its efficient operation. On the other hand, mounting the mill on the crest of a smooth hill will enhance the operation by providing higher wind velocities. Modification of the topography may also be helpful. Mills should be installed at least 500 feet from any obstacle, such as trees or houses.

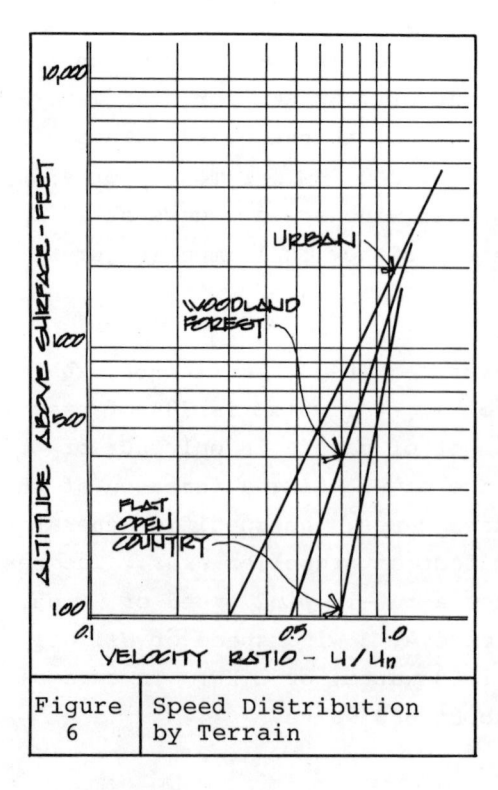

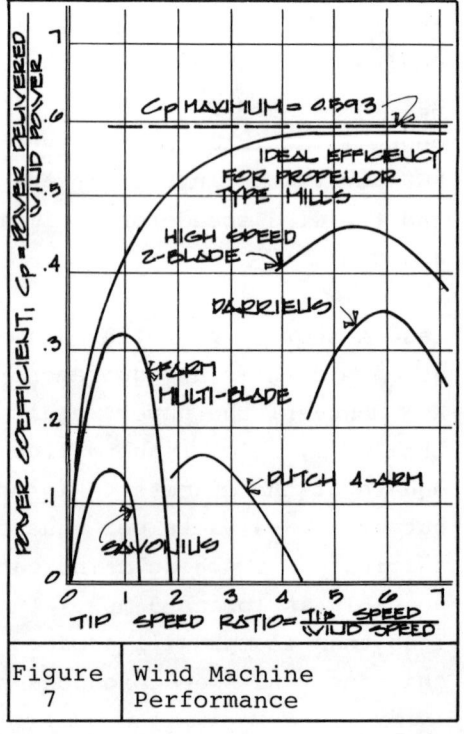

Figure 6	Speed Distribution by Terrain

Figure 7	Wind Machine Performance

Solar Architecture

The energy-storage system if used, is somewhat critical in so far as its sizing is concerned. Too small, and great inconvenience results; too large, and economy has been sacrificed. For battery storage, a special type of battery is used, known as houselighting or stationary batteries. These have thicker plates than automotive-type batteries and generally have a lifetime of about 2,000 complete cycles, from fully charged to fully discharged to fully charged.

The combined totals of power needs and storage system size required determine the generator size. Large propellers turn as slowly as 25 rpm and smaller ones at speeds up to 400 rpm, hence a low speed generator is necessary and it must be designed to match the torque and power output of the prop, or vice versa. For AC output, multi-pole alternators are used. Alternators are preferable to generators for two reasons: (1) For the same wattage output the alternator operates at a lower cost than the generator, and (2) the high current from the alternator is taken from the stator coils whereas in the DC generator the high current is taken from brushes, which eventually wear and require replacement. Conventional alternators generally are designed for 1800 to 3600 rpm, whereas the wind system alternators should be designed for rotational speeds of 600 to 800 rpm. This allows for a low gear ratio (alternator rpm/prop rpm) so that losses and wear may be minimized.

In a range of tip speed ratios (speed of turbine tip/wind speed) from three to ten, the most efficient turbine is the propeller type, having two, three, or four blades (Figures 7). This type yeilds a high lift to drag ratio, light construction, is easily governed by simple automatic mechanisms, and permits lightweight, compact reduction gearing. Static and dynamic balancing, however, are very critical and considerable care must be taken during construction so that excessive vibration does not result.

Two-bladed props do not operate as smoothly during wind direction changes as do three-bladed props, and for steady-state operation the three blades are more aerodynamically efficient than two blades at any given velocity ratio. Four-bladed systems operate smoothly during wind direction changes, have reasonably high efficiency and operate best at tip speed ratios from three to four. They are, of course, more expensive and heavier than two and three blade types.

The cost per installed KW probably will be somewhat higher for the wind generator than for a conventional installation. However, the cost of fuel for the wind generator is zero, whereas fuel costs for conventional power generation have been escalating rapidly, and even in the case of coal utilization, the fuel costs continue to increase. When the overall cost per KWH is considered, the wind-driven generator becomes a highly competitive method of electrical power generation (Figures 8 and 9).

CONCLUSION

In conclusion, wind-driven electrical generators can replace (or

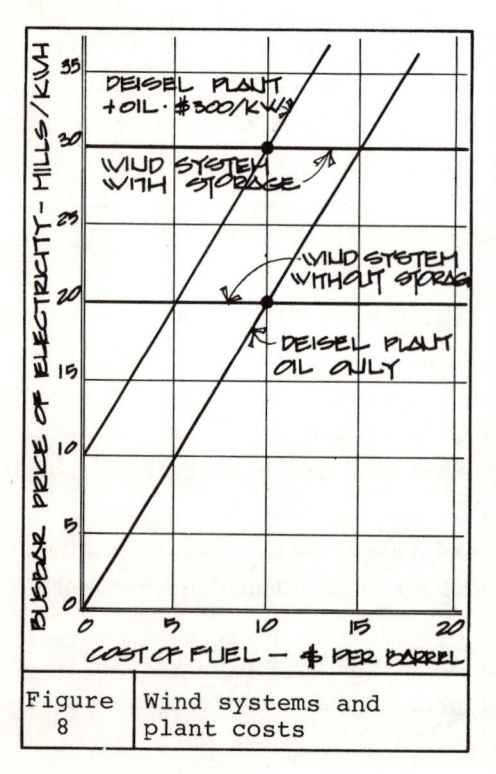

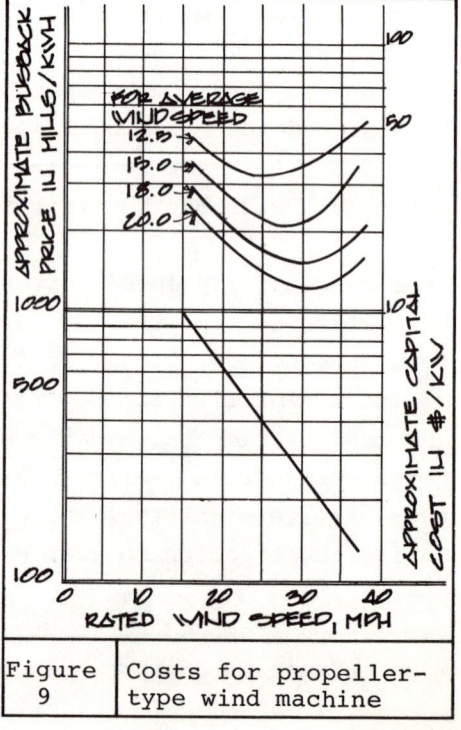

Figure 8	Wind systems and plant costs

Figure 9	Costs for propeller-type wind machine

add to) conventional fuel oil and natural gas drive generators for large portion of time, thus conserving large amounts of these fuels which are subject to short supply and high prices. It is estimated that about half of the fuel oil and natural gas used by utilities can be conserved over any 12-month period.

BIOGRAPHICAL SKETCH
Stan H. Lowy is a professor of aerospace engineering at Texas A & M University in College Station, Texas.

ENDNOTES
1. Abbott, I.H. and Von Downhoff, A.E., *Theory of Wing Section*, Dover Publishing Co., 1959.
2. Barnes, H.A. et al, *Wind Energy Power Production Systems*, Aerospace Engineering Dept., Texas A & M University, College Station, Texas, 1975.
3. Eldridge, F.R., *Wind Machines*, the Mitre Corporation, McLean, Virginia, 1975.
4. Golding, E.W., *The Generation of Electricity by Wind Power*, Philosophical Library, Inc., New York, 1955.
5. Hewson, E.W. et al, *Wind Power Potential in Selected Areas of Oregon*; Reports PUD 73-1, March 1973; PUD 74-2A, August 1974; PUD 75-3, August 1975, Oregon State University, Corvallis, Oregon.
6. Lowy, S.H., Approximate Stress Analysis for Guyed Towers, *Civil Engineering*, pp. 80-81, March 1972.
7. Mantell, C.L., *Batteries and Energy Systems*, McGraw-Hill Book Co., New York, 1948
8. Putnam, P.C., *Power from the Wind*, Van Nostrand Reinhold Co., New York, 1948.
9. Savino, J.M.(ed), *Wind Energy Conversion Systems*, Workshop Proceedings, June 11-13 1973, September 1975, Washington, DC.
10. Syverson, C.D. and Symons, J.G., *Wind Power*, J.G. Symons Jr., Mankato, Minnesota, 1973.
11. "Energy Primer," Portola Institute, Menlo Park, California, 1974.
12. Wind Energy Society of America, Newsletter No. 9, "A Direct Method for Selecting Wind Turbine Generator," P.G. Sulzer,

Sulzer Assoc., Pasadena California, 1975.
13. Lubrication, Vol. 61, October-December 1975, Texaco, Inc., New York.

WOOD FOR ENERGY

Presented and prepared by: Ja Densmore

This paper presents an informal discussion of wood burning and the changes that have taken place since the energy crises hit the East coast in the fall of 1973. Wood is a form of solar energy--it takes the sun energy to grow a tree. The difference is that Mother Nature found a way to store the energy until needed. It must be remembered that wood is a renewable resource. There is more wood rotting on the forest floors than is used in all the forest products in the country.

If all the woodlands in this country from coast to coast, were harvested and farmed correctly, it could supply 20% of the nation's energy without hurting the forests.[1] This was the statement of Mr. Tom Reed from the MIT Lincoln Labs when he addressed the Wood Energy Institute in Cambridge, Massachusetts in April, 1977. This statement, of course, means using all types of wood available for both private and commercial use-- this would include wood stoves, wood furnaces, woodchip burning and wood gasification.

WOOD BURNING

Wood burning has had many changes and improvements since late 1973 when the energy shortage hit the East Coast. There are very few homes in the Northeast that do not have some form of woodburner--either a fireplace or a wood stove. These units have been very helpful to the occupants in savings on their fuel oil and gas bills. In the fall of 1973, fuel oil cost 16¢ per gallon; now this price is up to 50¢. People just cannot afford to heat their homes in this manner. One man in New London, New Hampshire, lives high on a hill--the ridge pole of his house is 120 feet long. In January 1973, he used 8,000 KWH of electricity to heat his home. During the summer, he installed a Jøtul woodstove in the center of his house. In January 1976, which had more degree days than 1975, he used only 2,000 KWH--a saving of 6,000 KWH. He now has two other woodstoves, one at each end of the house. He uses these when the temperature is 10 degrees and falling. This is one way to beat the high cost of electricity.

In regard to fireplaces: the fireplaces that we have all known and grown up with will be a thing of the past in the very near future as they are at best only 5 to 15% efficient. NASA and ERDA have built a tech house at the Langley, Virginia testing facilities. This house is actively solar heated. Solar collectors are on the roof of the living modular and the bedroom modular. These are used for heat and domestic hot water. There is a solar cell on the roof of the garage for use of low voltage electricity; the only other source of heat is the fireplace. They have modified the construction of the conventional fireplace as follows: taking air from outside the house and piping it to the combustion box of the fireplace (this means that the warm air from the room is not used for combustion and thus wasted up the chimney). The firebox itself has a circulating unit around it so that cold air from the floor of the room can be heated and then circulated back into the room--the use of two low velocity fans placed in the ducts at the floor level can aid in this. The next change they made was to put glass doors on the front of the fireplace. These glass doors are available commercially from any complete fireplace store and should be installed on every existing fireplace. When you are burning a fireplace in the evening, it is impossible to close the damper when you go to bed

Solar Architecture 243

because there is still fire and hot coals left in the fireplace. The result is that warm room air is drawn up the chimney all night long. The glass door prevents this. The last modification that NASA made was to install water coils to the inside back of the fireplace so that when it is in use water is being heated and can be piped to the hot water storage.

PITKIN COUNTY LEGISLATION

In line with the above, the Board of Commissioners of Pitkin County, Colorado have passed an ordinance for air quality standards and included in this ordinance is a limitation on the number of fireplaces in new construction. This ordinance also states that: all fireplaces shall be constructed such that their operation will increase heat energy supplied to the living area in quantities greater than that lost through air exchange during combustion; and in addition, be constructed in conformance with any design standards that may be promulgated (or approved) by the County engineer which are designed to increase heat energy supplied.[2]

The term fireplace as used in the ordinance includes a conventional masonry fireplace, a prefabricated zero clearance fireplace, and any similar fireplace whose operation requires it to be built into the structure as a component of the building. Radiant room heaters, heating stoves, and similar appliances designed for space heating purposes are not included within the definition of "fireplace" and are not subject to the limitations set forth for fireplaces. In passing this ordinance, the County commissioners are allowing the residents of Pitkin County to utilize this form of energy saving which has been used so successfully in the East. For the thought which has gone into this ordinance, they should be commended.

WOOD STOVES

Many changes have taken place in the wood stove (i.e. the box stove, pot belly, parlor, and the Franklin). When the

energy crisis hit, these units were brought back into use once
again. The foundries still had the patterns and they produced these in great quantities. These stoves are made of
cast iron and perform quite well. However, the energy crisis
produced the study and the development of an efficient wood
heater in this country. The first to appear on the American
market were the very efficient Scandinavian stoves which had
been in use for many years. These stoves are air-tight, made
of heavy cast iron, and they have a pleasing appearance.
Since early 1974, a number of American manufacturers have produced efficient wood heaters. All of these wood heaters, both
Scandinavian and American, work on basically the same principle with slight modifications and methods of handling the same
principle of burning with wood.

When wood burns, it goes through three stages; in the first
stage, the free water is driven off; in the second stage, the
wood is broken down into charcoal; finally, in the third
stage, the charcoal itself burns producing most of the heat.
Combustion of the volatiles can deliver 60% of the heat potential in a log if the stove is designed to burn these gases.
In order for these gases to burn, they must reach approximately
1100F and they must be mixed with oxygen. It is difficult for
a wood stove to meet these conditions. A primary air supply
is needed to feed the lower part of the fire during the first
two stages of burning. However, during the third stage (the
burning of the charcoal), the oxygen is used up so that none
is left for the burning of the gases given off. Therefore, it
is necessary for the stove to furnish a secondary air supply
in just the correct amount and directed to the top of the fire
so that the gases will burn and produce heat for the room, not
just go up the chimney. The advantages of the airtight stove,
by achieving the full burn of wood, is that they will burn and
hold their heat for a long period of time. Many of them can
be charged with wood before going to bed and one will still
have a hot stove with red coals the next morning. This method
of wood burning makes it very advantageous in combating the
high cost of fossil fuels. The airtight stove is an effective

Solar Architecture

backup heat when used in conjunction with solar-heated homes. Manufacturers have different ideas on how to achieve this, and there are a number of stoves on the market today which do an excellent job.

Wood type
Many people ask the difference in using hard or soft wood in their stoves. In general, hardwoods last longer in a fire and will generate more coals. Softwoods burn faster and hotter. This is probably true in an open fireplace or loosely constructed woodstove. However, in an airtight stove, the rate of burn can be controlled by the amount of air allowed to the fire. In Norway, soft wood is used almost exclusively in the burning of their airtight stoves. To within a very few percent, a pound of wood of any kind, either soft or hard, has the same Btu content. The major difference between woods is their weights or densities. Wood should be dry when burned. Some people prefer hardwood over softwood. However, all types of wood are satisfactory as fuel. In practice, people burn whatever is most readily available.

Safety
Safety and common sense are two traits that should be used when burning with wood. Above all else, stove installations should be safe. By complying with the safety standards of the National Fire Protection Association in regard to minimum clearances from combustibles and the local building codes, virtually all danger is eliminated except operator carelessness.[3] The above is true for the initial installation, however standards do not guarantee safety. Stoves, stove pipe, connectors, and chimneys should have a periodic check and cleaning to remain safe. Creosote is one problem of the airtight stove. Wood combustion is never perfectly complete which means that some of the unburned gases and tar-like liquids can condense on the cool walls of the chimney. This formation is dark brown or black in color and is called cresote. All kinds of wood can form cresote. It is inevitable and must be lived with. This formation can cause chimney fires and usually occurs during

a very hot fire. The best way to eliminate a chimney fire is
to keep a clean chimney. In Norway all chimneys are swept
twice a year by government chimney sweeps. In this country,
the chimney sweepers have formed a guild and we shall be
seeing a lot more of them in the future. Many people sweep
their own chimneys. They do this by using a stiff wire brush
the same size as the inside of the chimney. If a person will
use dry wood and learn the burning characteristics of his
wood with his stove, he will do a lot to increase the heat
energy of his unit and also keep the formation of cresote to
a minimum.

CONCLUSION
A very interesting development in this day of alternative
heating methods should be noted. In 1977, the House of Representatives of the State of Maine have passed a bill which
will exempt certain woodburning appliances from the state
sales tax.[4] This act is designed to encourage the practice of
woodburning under the state policy of providing tax incentives
to develop the renewable natural resources of the state. This
act should be followed by other states.

"Nature has been growing and decaying trees for hundreds of
millions of years. The solar energy locked into chemical form
in the plants through photosynthesis has always been released
mostly as heat during decay. By taking the wood and burning
it in a stove, the heat is released in a home rather than on
the forest floor. The stored solar energy ends up in the same
form in either case; burning the wood merely reroutes the
energy through a house on its way back into the atmosphere and
eventually back into space. The amounts of carbon dioxide
and oxygen in the atmosphere are not significantly affected,
whether wood is burned or left to decay".[5]

BIOGRAPHICAL SKETCH
Ja Densmore has been actively working with the development
of woodburners for years and is the owner of the Burning
Log in Aspen, Colorado.

ENDNOTES

1. *Proceedings Document: Wood Heating Seminar I*, Wood Energy Institute, Box 1, Fiddlers Green, Waitsfield, Vermont.
2. Resolution No. 77, Board of County Commissioners of Pitkin County, Colorado.
3. *Manual of Clearances for Heat Producing Appliances*, NFPA No. 89 M 1971, National Fire Protection Association, Boston, Massachusetts.
4. Legislative Document No. 1465, State of Maine House of Representatives, April 7, 1977.
5. Shelton, Jay and Shapiro, Andrew, *The Woodburners Encyclopedia*, Vermont Crossroads Press, Inc., Waitsfield, Vermont.

THE MEANING AND APPLICATION OF GEOTHERMAL ENERGY

Presented and prepared by: Glenn E. Coury

Geothermal energy literally refers to the heat of the earth. It is being used as a source of heat energy and as the fuel for electricity generation. High heat sources can be found anywhere on earth by simply drilling deeply enough. However, to be useful, the heat source must lie in an economic deposit. That means, it must be reasonably close to the surface (less than 10,000 feet) and at a high temperature (greater than 400F for electric power production) and it must be stored in a useful form, as in hot water or steam (hot, dry rocks located at depth cannot now be used economically). As the technology develops, however, all of these constraints can be expected to loosen.

Geothermal energy is an attractive concept that has been discussed and analyzed at exponentially increasing rates during the past six years, and it truly represents a potentially vast energy resource. Utilization of the resource has not yet been rapid, however, for both technical and non-technical reasons. Non-technical deterrents relate to such factors as leasing government lands, environmental impact statements, tax laws, mining laws and, of course, economics. Technical problems are related to techniques for finding and determining the size and quality of geothermal deposits, producing brines that are often corrosive and chemically scaling, and developing economic processes to convert the energy contained in low-temperature brines to a more useful form.

Nevertheless, a firm commitment to this new industry has been made by industry and the government. Numerous leases on public and private land have been taken by private companies and many wells have been drilled. The U.S. Energy Research and Development Administration (ERDA) has an annual budget approaching 100 million dollars to support the development of the industry. As the economics and reliability of the industry become proven, it will grow inexorably.

The largest current utilization of geothermal energy is for the production of electricity. For example, dry steam at 300 to 400 F produced from wells of depths ranging from 2000 to 10,000 feet, drives turbine generator sets in the United States, Italy and Japan. Hot brines at underground temperatures of 500 to 700F are flashed up deep wells in New Zealand and Mexico; the steam fraction is sent to power plants while the residual hot brine is discarded. Lower temperature waters of up to 200 F are used in many parts of the world for space heating of homes, greenhouses and commercial buildings.

The bulk of the economically significant geothermal deposits is expected to exist in the form of moderate temperature brines in the range of 300 to 500 F. For these brines, the economics of production, as related to both the size of the resevoir and the productivity of a single well, will be perhaps the most critical factor with respect to the overall profitability of a project. This can be illustrated by comparing the energy content of hot water to that of oil. Oil contains about 5.6 million Btu per barrel, which can be released during combustion at high temperatures to provide for the relatively efficient production of electricity. One pound of water at 500 F, however, contains only about 350 Btu of usable energy which, because of the low temperature, is used with only a low efficiency when converted to electricity. Thus, for the same output of useful energy, 100 to 200 times more brine than oil must be produced.

Solar Architecture

Larger wells must be used and, to keep well costs from being prohibitive, the production rate per well must be very large for a low temperature source to be economic for the production of electricity.

On the average, the earth temperature will increase 20 to 30°C per kilometer of depth into the earth. There is a small depth near the surface of the earth where the temperatures vary considerably due to the daily ambient cycles. The earth has a crust on it, and below the crust is a mantel. The crust is usually 25-50 kilometers in depth. It varies depending on location. At the bottom of the crust, 1,000°C temperatures are expected. The core of the earth is about 4,500°C. There are not many industrial processes that work at that high of temperature. Most furnaces will burn gas or coal at around 1,100°C, and some of these processes will work at higher temperatures. Practically all energy utilization is at temperatures much less than that. The temperatures for making electricity from burning coal or gas are much lower than 1,000°C. So, the earth certainly has temperatures that are hot enough. The minimum temperature to make electricity is 200°C. And using the normal temperature gradient in the earth, one would have to go seven to eight kilometers (maybe 20,000 feet) to reach that temperature.

Geothermal energy is being generated all of the time. One of the theories about generation of heat is friction. The crustal places of the earth and the mantel are actually sliding and moving. Great plates meet each other, grind and crunch and thus generate a lot of heat by friction. There is also an enormous amount of radioactive material that is very diluted; not concentrated enough for economical mining of uranium or other similar materials. But each one of these radioactive particles is generating a little bit of heat that results in a large amount of heat being generated continually in the earth. As this heat escapes to the surface, it creates a temperature gradient which is the source of geothermal energy.

In terms of the quantity of heat now being stored (in the top kilometers of the earth's crust), it is in the order of 10^{26} power calories, and to put that into perspective the worldwide consumption of energy is of the 10^{26} calories. That means there is enough stored right now in the top ten kilometers to last for a million years. (That is, if it could be extracted out of the crust.) The problem is getting this 10^{26} calories out of the top ten kilometers of the earth's crust in an economical way.

A lot of big companies are putting up money to explore and find geothermal sources. The main reason that geothermal energy has developed slowly is that the most desirable land for geothermal development belongs to the government. It was only a few years ago that the federal government started leasing land for geothermal energy. There is a whole maze of leasing, legal, regulatory and environmental activity that has to be completed before one can initiate the development of land. It was in late '71 that the law was passed allowing lands to be leased for geothermal energy. It took three years for the regulations to be written by which this law would be implemented and the land leased. And since then there have been a number of these lease sales. Of course, there have been lease sales on private lands. There is a major development in Northern California at the Geysers which has been primarily on private lands.

Once a geothermal source has been discovered, the engineering technology exists to use it. Sometimes the problems are very difficult. But problems are always difficult in any large manufacturing or producing area. The nature of the problems are due to the fact that the hot water that comes from underground contains a lot of dissolved salts. These salts are corrosive, and as they go through the pipes they deposit salts and plug lines, resulting in a very costly activity. These are problems that must be resolved and they can be, at a cost. Most of them can be resolved at a reasonable cost. Other problems are the dissolved gases that are contained in geothermal water or geothermal steam. Some of these gases are poisonous and can be fatal.

In other countries, low-temperature brines have been very effectively utilized on a large scale for heating homes, businesses and greenhouses. Iceland is the most notable example; where most of the homes in the City of Reykjavik are so heated. In parts of Russia and Hungary, large acreages of greenhouses are geothermally heated. Even a suburb of Paris, France finds an apartment complex heated by natural hot waters, and similar heating projects have been established in Klamath Falls, Oregon and Boise, Idaho.

In brief, the technology is available for the utilization of geothermal waters and the economics are often favorable. Its widespread development hinges largely on finding and developing economic deposits, much as economic deposits of oil are found in locations that are situated near the ultimate market.

BIOGRAPHICAL SKETCH
Glenn E. Coury is a geothermal consultant and chemical engineer with Coury and Associates, Inc. of Lakewood, Colorado.

ENERGY INDEPENDENCE THROUGH METHANE

Presented by: C.E. Tomson, Jr.
Prepared by: W.W. Marlatt, C.E. Tomson, Jr.

INTRODUCTION

Methane production from organic waste material is the cheapest and most efficient source of heat energy. Over a 20-year period methane would cost between $.70 and $1.26 per million Btu, depending upon the digester size. To buy an equivalent amount of propane at $.35 per gallon would cost $3.83 per million Btu. Natural gas priced at $2.50 per thousand cubic foot would cost $2.62 per million Btu. Solar energy estimates are $5 per million Btu.

The efficiency of methane production will range from 90 to 96%. This means it only takes 4 to 10% of the fuel produced to keep the digestion process functioning. Solar efficiencies are approximately 60 to 65% of energy received per energy collected. Wind energy converts to electricity at a 33% efficiency, while geothermal conversion to electricity is between 30 to 40%.

Organic material for methane digesters is abundant. The primary sources are manure from livestock confinement centers, raw sewage and garbage. Many municipal waste treatment facilities employ anaerobic methods as one of the steps in processing raw sewage. Here the emphasis has been largely on the destruction or digestion of the sewage, and only recently have some cities focused on the recovery and use of the methane gas produced.

Estimates show that in 1970 two billion tons of fresh animal manure were produced in the United States of which approximately 25% is readily collectible. If this 25% were converted to methane gas, it would yield approximately 1.4 trillion cubic feet of gas per year. This amount is about 7% of all natural gas used by this country in 1970.[1] The anaerobic bioconversion of all available animal manure, sewage, garbage, etc. will not make the entire United States energy self-sufficient. However, most livestock enterprises and many home owners could become energy self-sufficient through the bioconversion of their individual supplies of organic raw material.

HISTORY OF METHANE

The history of sludge digestion and its precursors can be traced from the 1850's with the development of the first tank designed to separate and retain solids. The first unit used to treat settled sewage solids was known as the Mouras automatic scavenger, which was developed by Louis H. Mouras of Vesoul, France, in about 1860 after it had been observed that if the solids were kept in a closed valut (cesspool) they were converted to a liquid state. The first to recognize that a combustible gas containing methane was produced when sewage solids were liquefied was Donald Cameron, who built the first septic tank for the city of Exeter, England in 1895. He collected and used the gas for lighting in the vicinity of the plant. In 1904 the first dual-purpose tank incorporating sedimentation and sludge treatment was installed at Hampton, England. It was known as the Travis hydrolytic tank and continued in operation until 1936. Experiments on a similar unit, called a Biolytic tank, were carried out in the United States between 1909 and 1912. In 1904 a patent was issued to Dr. Karl Imhoff in Germany for a dual-purpose tank now commonly known as the Imhoff tank. One of the first installations in the United States employing separate digestion tanks was the sewage treatment plant in Baltimore, Maryland. Three rectangular digestion tanks were built as part of the original plant in 1914, and an additional tank

Solar Architecture
was added in 1921.[2]

Since about 1930 many individuals in India and Europe have attempted to develop practical and efficient systems to produce methane from organic wastes. The fuel shortages during World War II prompted Germany to expend much effort toward the experimental development of actual gas-producing plants for large farm and animal production centers. Manure was abundant and was bioconverted to produce the gas. A number of designs were developed, the main ones being the Schmidt-Eggergluss, the Weber and the Kronseder types. In France by 1952 there were approximately 1,000 installations. In India by 1973 there were more than 2,500 small installations producing methane. India has projected that by 1979 there will be 100,000 methane digesters in operation.[3]

THE BIOLOGY OF DIGESTION
Anaerobic Digestion Process
Swamp gas or marsh gas is approximately 60% methane produced by bacteria acting on submerged organic material in stagnant bodies of water, where the oxygen content is very low. Another example is the gas produced by digestion of food within the rumen and large intestine of warm-blooded animals. The exclusion of oxygen enables certain species of bacteria to flourish with the resultant production of methane-rich by-product gases.

This natural anaerobic decaying process can be controlled by placing organic matter into an air-tight container, then maintaining optimal temperature, pH, and feeding rates for the bacteria of decomposition. It is very important to remember that this decomposition is a <u>biological</u> process involving a series of reactions by two types of anaerobic bacteria. The first type of bacteria are called acid-producers, and it is this group which initially decomposes the organic matter by the action of extracellular enzymes. Thus, the large complex molecules (carbohydrates, proteins and fats) of the organic matter are reduced to smaller simple compounds (simple sugars, amino acids and volatile fatty acids) which can be absorbed through the cell walls of the acid-forming bacteria, thereby entering into their metabolic processes. The end products of this phase of bacterial hydrolysis and metabolism consist of carbon dioxide, small amounts of hydrogen and relatively large quantities of short-chain volatile fatty acids. Acetic, proprionic and butyric acids form the bulk of the volatile fatty acids produced, with acetic acid being responsible for about 70% of the ultimate methane production.[4]

The next phase of bacterial fermentation consists of the volatile fatty acids being metabolized by a second group of bacteria. The original organic matter fed is almost totally converted by bacterial action into bacterial protoplasm, certain dissolved by-products and gaseous end products (carbon dioxide and methane).

One group of methogenic bacteria function at temperatures ranging from 60 F to 110 F with 90 F being the optimum point for methane production. This is known as the mesophilic range. Another group of methogenic bacteria species function at higher temperatures ranging from 120 F to 150 F; the optimum point being 130 F. This is the thermophilic range. Whichever temperature range is chosen, a constant temperature is important since a rise or fall of only 3 F to 5 F will inhibit the sensitive methogenic bacteria.[5] Thermophilic methogenic bacteria have much higher metabolic rates and, therefore, digest and produce gas at a faster rate. However, they are much more sensitive to variations in their environment, and more of the produced gas must be used to maintain the higher digester temperatures.

The methogenic bacteria generally function in the pH range of 6.4 to 7.5 with 7.0 or neutral pH being the optimum point.[6] The pH level is controlled by the organic feedstock loading rate. If organic matter is added too rapidly the acid-forming bacteria, which are quite hardy and multiply rapidly, will produce a surplus of volatile acids, thereby lowering the pH from the optimum range for the methogenic bacteria. High pH or alkaline conditions in a digester are most often caused by excessive concentrations of un-ionized ammonia. This results from feeding the digester with organic feedstock containing a deficiency of carbon relative to the amount of nitrogen present. In practice, a properly sized digester that is fed the calculated load of organic feedstock on a regular basis will seldom have severe pH fluctuations. This is due to the fact that buffer systems become established that are able to handle minor day-to-day variation in feeding quantity or quality.

Digestible Raw Materials
Animal manure, raw sewage, garbage, blood, grass clippings, wheat stubble, rags, leaves, corn stover and sawdust can be fed into an anaerobic digester. Whatever the feedstock though, analyses must be obtained to determine the amount of nutrients that will be available for bacterial action. An effective method of determining this value is to oven dry a sample of potential feedstock at 220F until no more weight is lost from dehydration. The sample is then called Total Solids (TS). It is weighed, then returned to the oven and heated at approximately 1000F which volatilizes and drives off all organic compounds present. This residue is now weighed and is called Fixed Solids (FS). This is the biologically inert portion of the sample. The Fixed Solids are then subtracted from the Total Solids with the resultant portion being called the Volatile Solids (VS). This portion represents the amount of true organic matter available to the bacteria. However, a small amount of the VS is usually non-digestible or only very slowly digestible. This portion generally consists of plant fibers such as lignin or animal parts such as hair or feathers. Nevertheless, the VS content of an organic feedstock is a very helpful value to know and is usually the basis for calculating gas production, loading rates and retention time.

Another very important parameter to consider in a potential organic feedstock is the ratio of carbon to nitrogen (C/N). For maximum methane production, the C/N ratio should be about 30 to 1. Organic material with higher C/N ratios than this will produce a raw gas with less methane and more CO_2, while C/N ratios substantially below 30 to 1 will produce danger of ammonia toxicity with resultant high pH and methogenic bacterial inhibition.[7]

Manure from cattle, poultry and hogs has generally been considered the most practical organic material for large-scale anaerobic digestion, since it has a relatively high volatile solids concentration and favorable C/N ratios. It has the added advantage of often being available in large quantities due to concentrated livestock populations such as feedlots and confinement buildings. Animal waste generally consists of the following substances:
1. Undigested or partially digested food
2. Digestive juices
3. Biological products of metabolism
4. Living or dead bacteria from the digestive tract
5. Cells and cell debris from the digestive tract wall
6. Spilled feed
7. Animal hair or feathers
8. Water
9. Bedding or litter (if used)
10. Soil or sand (if from outdoor pens or earth floors)
11. Milking center wastes (if dairy)

Table 1 illustrates the species differences for animals' manures when analyzed for TS, VS and C/N ratio. It can be seen that the C/N ratio of poultry is quite low due to the inclusion of nitrogeneous urinary wastes with digestive tract wastes in avian species. In this case, full methane production from poultry manure can be obtained only through the addition of some digestible material with a much higher C/N ratio in order to obtain a mixture with a C/N ratio approaching optimal levels. The values of Table 1 are only general averages, and much variation is found in individual sources of manure. Some of the major factors affecting manure quality are the feed ration of the animals, the manure age and, if not fresh, the climatic conditions during its storage. For instance, one-year-old manure will have undergone considerable aerobic decomposition and possible leeching of nutrients due to precipitation; therefore, the VS content would be lowered and the C/N ratio could be changed appreciably.

		% Moist	WET MANURE					TOTAL SOLIDS		VOLATILE SOLIDS	C/N RATIO
	lbs.		Lbs. Per Day	Gals. Per Day	Ft.^{3} Per Day	Tons Per Year	Lbs. Per Day	Tons Per Year	Lbs. Per Day	
LIVESTOCK										
Dairy Cow	1300	80	107	15	2.0	19.5	13.5	2.5	11.2	25
	1600	80	132	18	2.4	24	16.6	3.1	13.8	25
Dairy Heifer	1000	80	-85	11.2	1.5	15.5	9.2	1.7		
Beef Steer	500	80	45	5.2	0.7	8.2	5.8	1.0	5.9	
	1000	80	60	7.5	1.0	11.0	6.9	1.3	7.5	25
Horse	1000	75	45	6.7	0.9	8.2	9.4	1.7	1.6	
Hog (sow)	500		25	3.0	0.4	4.6	2.2	0.4	0.48	
Hog Feeder	100	82	6.5	1.1	.15	1.2	0.6	0.11	0.96	
	200	82	13	2.2	0.3	2.4	1.2	0.22	0.85	20
Sheep Feeder	100	68	4.0	.75	0.1	0.73	1.0	0.18	0.038	5
Hen (laying)	4	50-70	.21	.024	.003	.038	.054	0.01	0.048	5
(broiler)	4	50-70	.28	.029	.004	.051	.068	.012		
Turkey		75								3
Blood										
Sawdust										511
Grass Clippings		65								19
Straw/Oats										83
Straw/Wheat										12
Maize/Stalk/Leaves										52
Rags		10								12
Leaves		50								203

Table 1   Manure production of various livestock and organic materials

VALUE OF DIGESTION PRODUCTS

Methane Gas Description

Methane--an odorless, colorless, flammable gas--is the simplest possible molecule of the alkane or saturated hydrocarbon series. It consists of a single carbon atom surrounded by four hydrogen atoms. It is one of the lightest of pure gases, and at sea level at 77F will produce 978 Btu/cf when burned in excess air. Mixtures of 5.5 to 14% methane in air are explosive. Concentrations of more than 14% methane in air are flammable but not explosive.[7] Methane burns with a pale, faintly luminous flame (the yellow color of burning natural gas is produced by the small amounts of propane present). The combustion of methane yields carbon dioxide and water vapor making it a virutally non-polluting fuel. It has an octane rating of 120, making a very good fuel for use in stationary high compression internal combustion engines whether gasoline or diesel design.[8] For most purposes, it is not practical to use methane as a fuel for vehicles as the Btu content per cubic foot is rather low, and several heavy, high-pressure storage bottles would be required to provide as much travel distance per tank filling as conventional gasoline, diesel or propane fuels. Unlike propane and butane, methane is not easily liquefied. Its boiling point is -161.5°C (-258 F). In order to liquefy methane it must be cooled to at least -82.3°C (-116 F) while compressing it to 675 psi Above this temperature, methane cannot be liquefied by any amount of pressure.[7]

Methane can replace natural gas, propane or butane for any domestic or industrial applications. This includes all stationary gas or diesel engines (which operate pumps, air compressors or electric generators), gas cooking, heating, gas lighting, gas refrigeration systems, incubators, crop driers, etc.[8] When sufficient quantities of synthetic natural gas (SNG) are produced, it can be sold to commercial natural gas pipeline companies generally at a "new gas" interstate price. Because of federal regulations the intra-state price is lower.

In some instances (e.g., boiler heating or a very low--less than

Solar Architecture

5 ppm--H_2S concentration) it is preferable to eliminate the CO_2 and/or H_2S scrubbers. The resultant 30 to 40% CO_2 gas has a lower Btu content per cubic foot, but the CO_2 aids in the heat transfer through its high specific heat. Larger burner orifices and carburetor jets are necessary to admit sufficient gas for proper fuel combustion. When the gas is of SNG quality, no changes are required for domestic use. Burner jets of water heaters, cook stoves, furnaces, etc. will not have to be altered from natural gas use.

Single Cell Protein Description

A high quality, organic fertilizer is produced from anaerobic digestion. It is the organic residue effluent remaining after bacterial fermentation has been completed. This effluent has been called Single Cell Protein Biomass or SCP. It is composed of expired bacteria bodies, undigested or partially digested organic feedstock and all of the original feedstock Fixed Solids. This SCP can no longer be classified as manure, since it has undergone biotransformation. Analysis of the SCP indicates that it contains almost twice the concentrations per pound of nitrogen, phosphorous, potassium and trace minerals, as the manure which was originally fed into the digester.[9] This is because these elements were not consumed or removed from the digester with the gas but were concentrated by the removal of carbon, hydrogen and oxygen, thereby reducing the solids to one-half the original volume. If the solids retention time is greater than 20 days, the SCP presents none of the offensive odor of fly attraction properties of manure. It can be applied directly as a 90% liquid or dried to a fluffy material and spread evenly over the field. Because most of the nitrogen has been converted to stable protein in the bacterial cell walls, losses due to ammonia release are much lower than composted manure fertilizer. The nitrogen is released gradually over a three or four-year span, thus reducing annual fertilization requirements and increasing the organic matter percentage of the soil.[10]

Another use for this SCP residue has been proposed.[9] The crude protein concentration has doubled from the crude protein levels of the manure entering the digester, while at the same time the amino acid concentration has increased four times. This indicates a significant conversion of non-protein nitrogen into amino acids. The quantities of the various amino acids compare very favorably with the amino acid analysis of soybean or cottonseed meal. Thus, it has been suggested that the SCP residue could replace soybean or cottonseed meal as a protein supplement in animals' feed rations. It is also logical that the minerals in the diet could be recycled. Further work remains here in the area of palatability tests, feeding and rate of gain tests and carcass yields; however, this application looks very promising.

ANAEROBIC DIGESTER PLANT SYSTEMS
There are two major categories of anaerobic digester systems: the batch digester and the continuous displacement digester. The batch digester is loaded all at once and held at proper physical and chemical levels until all the gas is produced. At this time, the entire contents are removed and another batch is loaded. The advantages of this approach are largely that the most complete digestion is accomplished; therefore, the ultimate in total gas production from each unit of organic feedstock is obtained and also the SCP residue is of the highest possible quality. The main disadvantage is that the gas production begins slowly, rises to a peak of production in about two to three weeks, then gradually falls off for about a month until no more gas is being produced. Obviously, no gas is produced during the emptying and loading stages. This problem can be minimized by installing two or more batch digesters and staggering their cycles.

Virtually all efforts toward commercial production of methane from organic wastes is now being directed toward the continuous displacement digestion system. In this system, the digester is filled with water and a partial loading of manure or other organic feedstock, then maintained at the proper physical and chemical levels. Once fermentation has begun, small or slow continuous daily loadings are performed. These digesters generally have the shape of a large horizontal pipe. Loading is done at one end with each load displacing the previous load toward

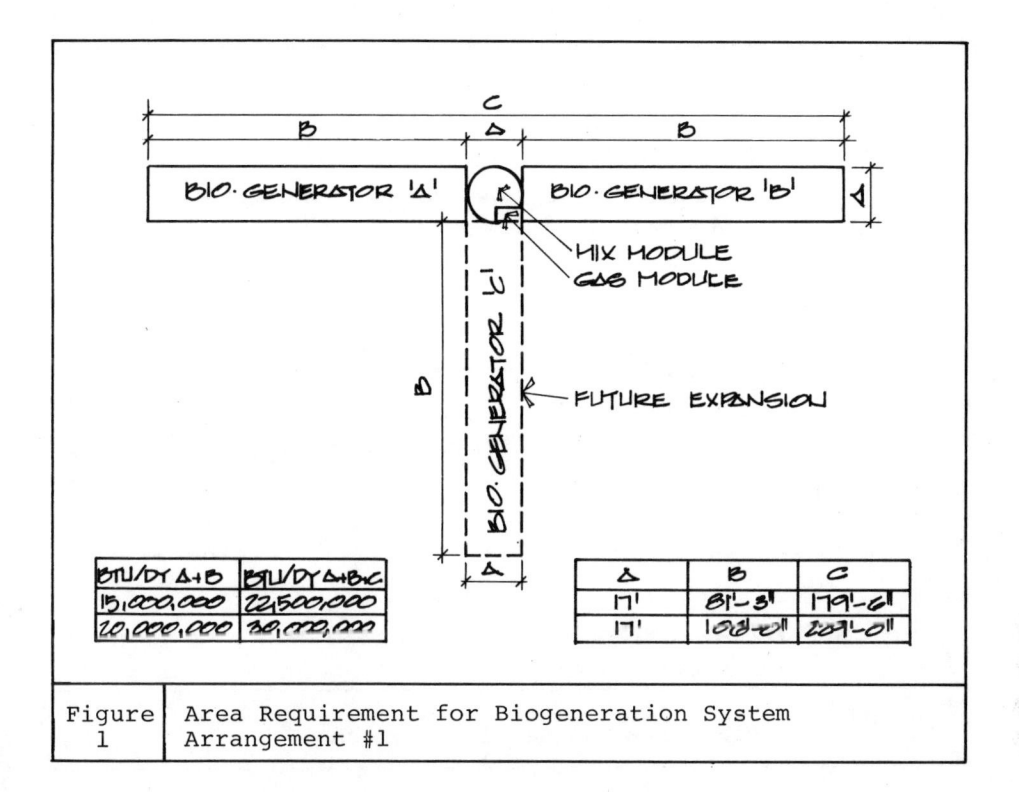

BTU/DY A+B	BTU/DY A+B+C
15,000,000	22,500,000
20,000,000	30,000,000

A	B	C
17'	81'-3"	179'-6"
17'	108'-0"	209'-0"

Figure 1	Area Requirement for Biogeneration System Arrangement #1

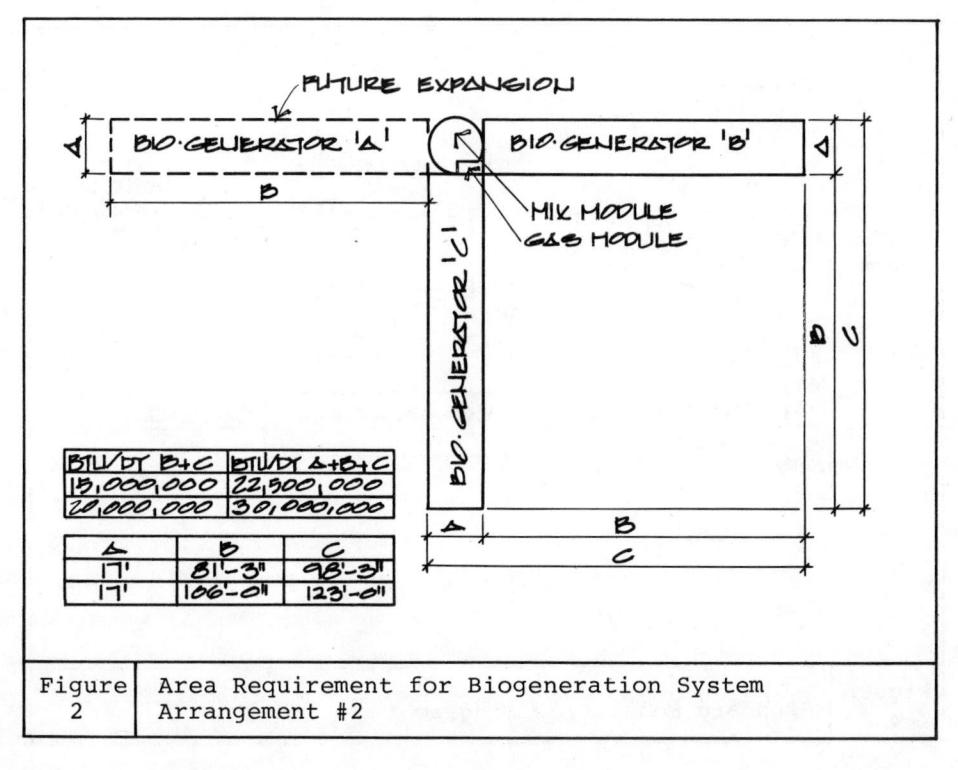

BTU/DY B+C	BTU/DY A+B+C
15,000,000	22,500,000
20,000,000	30,000,000

A	B	C
17'	81'-3"	98'-3"
17'	106'-0"	123'-0"

Figure 2	Area Requirement for Biogeneration System Arrangement #2

Figure 3. Flow Diagram - SNG Handling System

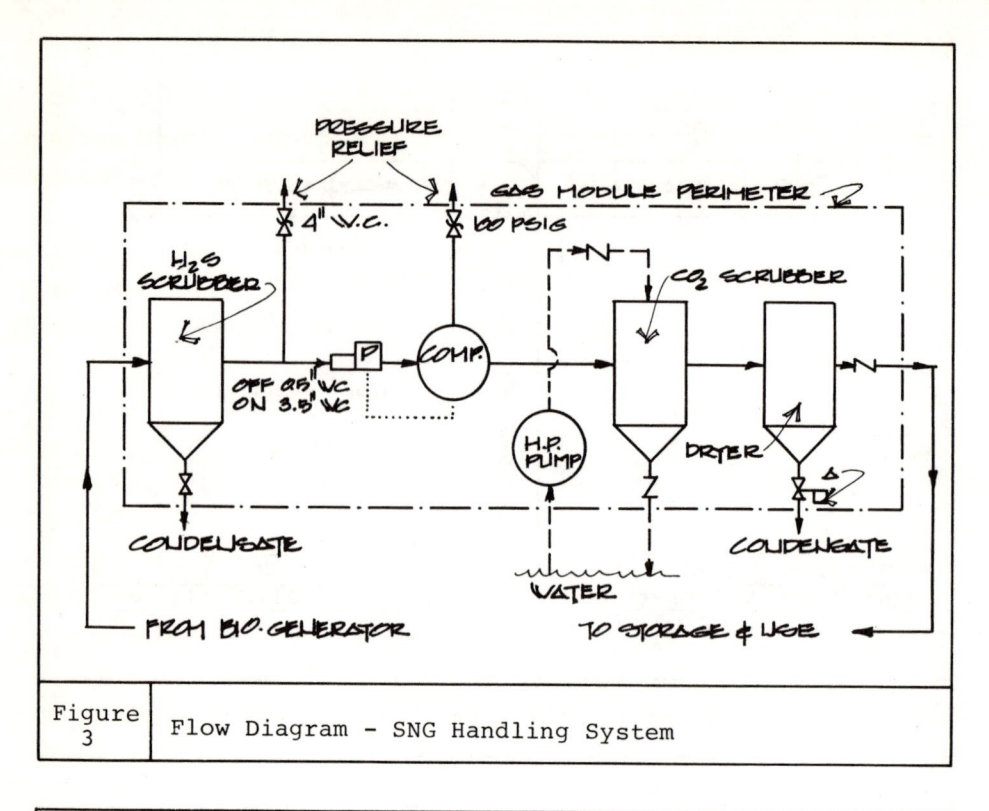

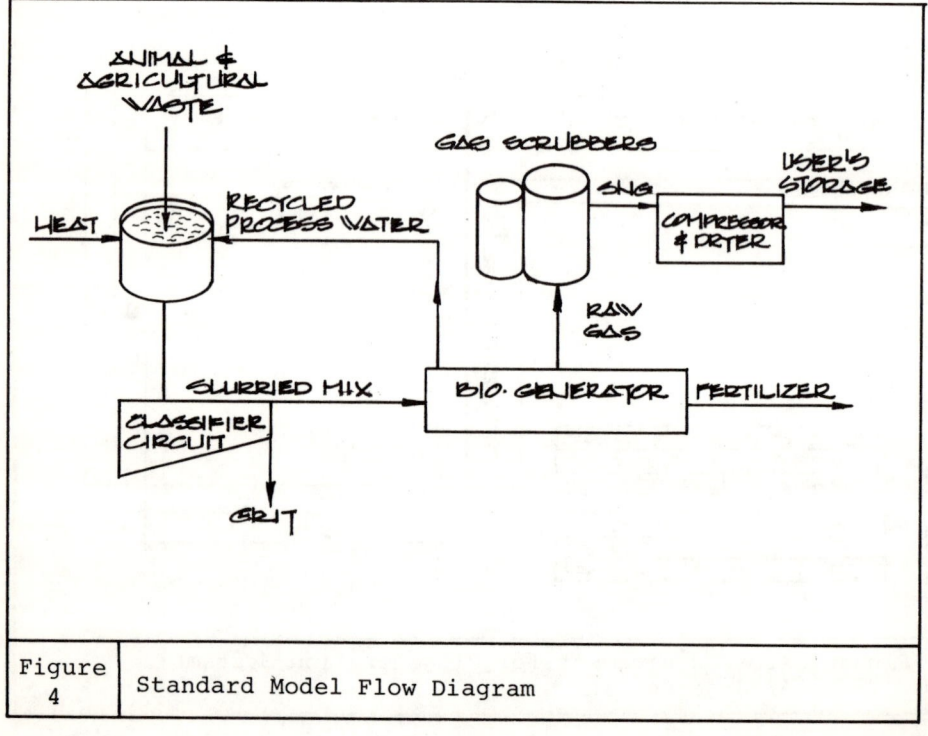

Figure 4. Standard Model Flow Diagram

Solar Architecture

the opposite end where, after a retention time of 20 to 30 days, the finished SCP residue accumulates and can be removed without shutting down or completely emptying the digester. The disadvantages here are largely that digestion does not have time to be totally completed. Gas production from a given unit of feedstock here is 90 to 95% of the ultimate that might be obtained from a batch digester. However, the advantages are thought to outweigh the loss of this final gas production. These advantages are that a constant supply of gas is obtained on a day-to-day basis, thus reducing storage requirements with no peaks to handle and no low or zero production from the unloading and loading stages.

Some additional equipment complications are encountered, since the digester is intended never to be opened once it is functioning. For this reason, provisions must be made to prevent or remove the SCP residue and other inert substances periodically. Sand and grit must be removed from the organic feedstock prior to loading to prevent settling out and accumulating within the digester. The continuous displacement digester system does, however, offer the best conditions for stable "production-line" operations.

A Specific Bio-Generation System Design

A 2,500,000 Btu Bio-Generation System (BGS) was installed at the Jim Miller Feedlot near LaSalle, Colorado in the spring of 1977. The unit was designed by Energy Recovery, Inc. of Broomfield, Colorado with construction by Agricultural Energy Systems, Inc. (AG-GAS) of Fort Collins, Colorado. The BGS is a continuous displacement type and is designed to operate in the mesophilic temperature range at 95 to 98 F. The system is designed to handle the manure from 63 feedlot steers, yielding 3,780 pounds of wet manure per day which contains approximately 500 pounds of VS. The holding capacity of the system is 18,000 to 20,000 gallons of manure slurry.

The BGS is composed of three basic modules: (1) mix module; (2) bio-generator vessel; and (3) gas/control module. The mix module consists of a 6 ft diameter by 10 ft tall, round aluminum tank which is covered and insulated. An SCP residue removal pipe with gate valve, heat exchanger, manure pump with necessary piping and a sediment classifier are incorporated. The bio-generator module consists of: (1) polyethylene moisture barriers; (2) styrofoam insulation; (3) aluminum roof assembly; (4) two vertical end walls; (5) thick-walled PVC bladder; and (6) gas/agitation system. The gas module consists of a small metal building enclosing a hot water heater, two water pumps, a 100 psi-rated gas compressor, water vapor dryer, CO_2 scrubber and H_2S scrubber. The automatic control circuits for the system are mounted on the outside wall of the gas module building. The BGS operates in the following sequence:

1. The cover on the mix module is opened and manure and/or crop wastes are dumped into the mix tank. The cover is then closed.

2. The start button is pushed. No further attention is necessary. The mix-transfer pump begins mixing the manure with liquid from the bio-generator module. During the mixing operation (approximately 30 minutes) the pump discharge is directed to the "degriter" where sand and gravel are removed. After completion of the mixing operation, the feed slurry is automatically pumped to the far end of the bio-generator vessel in which the actual conversion to gas and fertilizer occurs. The mix tank refills by gravity with liquid from the bio-generator vessel for the next day's use.

3. Gas evolving at a fairly uniform rate throughout the day fills the collection zone in the bio-generator vessel. Automatic pressure controls start and stop the gas module. The raw gas passes through a scrubber which traps the sulfur in the hydrogen sulfide thus freeing the hydrogen. The raw gas less hydrogen sulfide is compressed before entering the second scrubber which removes part of the carbon dioxide. Next, the gas passes through a dryer to remove the water vapor. The finished gas is synthetic natural gas of pipeline quality ready for direct use or storage. The use of a gas storage tank is recommended as a resevoir to meet peak usage, as gas is produced at a

Solar Architecture

constant rate throughout the day.

POSSIBLE FUTURE APPLICATIONS FOR METHANE

The potentials for anaerobic bioconversion systems of the future are immense. The most obvious applications of anaerobic bioconversion system products have been discussed previously. There are several other possible uses for the SNG, and there are other by-products that could be collected and utilized. Most of these are relatively high technology processes, but if large enough quantities of SNG are being produced these processes are possible and may prove to be economically viable alternatives.

Methane can be the *starting* ingredient for many chemical and biological conversions. Methanol, acetylene and halogenated hydrocarbons are just a few of the valuable chemicals that could be produced at the feedlot from the methane gas. A process has been developed and patented in which a mixed culture of micro-organisms are capable of oxidizing methane in a nutrient medium containing only inorganic salts to produce cellular material high in protein and vitamin content.[11] The cellular material produced can be used as a food or feed supplement for man or animals. Algae cultures can be grown in the supernatant liquid from the digester vessel. They will remove and utilize various salts and dissolved inorganic nitrogen and in turn produce a feed crop high in protein.[12]

Large amounts of CO_2 are produced during anaerobic bioconversion. This can be collected and purified, providing another by-product with a high value. It can be used or marketed as either compressed CO_2 or as dry ice. Approximately 1.2 tons of dry ice could be produced per day from the digestion of the manure of 1000 head of feedlot cattle. When the influent to a digester has an unavoidably high C/N ration (e.g. chicken manure), excessive amounts of ammonia are produced. It is possible to isolate and process this gas for use or sale as anhydrous ammonia fertilizer.

In summary, we have seen how methane obtained through the anaerobic bioconversion of organic matter can result in, or contribute toward, energy independence, especially on an individual basis. It is now possible, and indeed necessary, for individual homeowners, communities, ranchers and farmers to guarantee their supplies of fuel and energy. When adequate amounts of organic raw materials are available, methane production can be one of the most economical and reliable sources of alternative energy.

BIOGRAPHICAL SKETCH

Ed Tomson is the President of AG-GAS of Fort Collins, Colorado. AG-GAS designs and installs methane digesters.

ENDNOTES

1. Anderson, L., Energy Potential from Organic Wastes: A Review of Quantities and Sources, Bureau of Mines Info. Circular 8549, US Dept. of Interior, 1972.
2. Metcalf & Eddy, Wastewater Engineering, McGraw-Hill, 1972.
3. Singh, R.B., Bio-Gas and Its Place in Today's World Preserving the Environment, pp. 8-9, 1971.
4. Jeris, J. & McCarty, P., The Biochemistry of Methane Fermentation Using C^{14} Traces, J Water Poll, Control Fed., 37(2):178-192.
5. Boswell, A.M., Microbiology and Theory of Anaerobic Digestion, Sewage Works Journ, 19:28, 1947.
6. Sanders, F.A. & Bloodgood, D., The Effect of Nitrogen to Carbon Ratio on Anaerobic Decomposition, Water Poll, 37:1741, 1965.
7. The Merck Index of Chemicals and Drugs, 7th ed, Marck & Co.
8. Merrill, R., "Methane Reviews," Energy Primer, p. 144, 1974.
9. Coe, W.B., & Turk, M.,"Processing Animal Waste by Anaerobic Fermentation," Proceedings of the 33rd Annual Conference, Waste Utilization for Environmental Quality & Profit," 1972.
10. Hart, S., Sludge Digestion Tests of Livestock Manures, Dept. of Agricultural Engineering, Univ. of Calif. Davis, 1960.
11. Gutcho, S., Proteins from Hydrocarbons, Noyes Data Corporation: Park Ridge, NJ, pp. 47-48, 1960.
12. Burford, J.L., Jr. & Varani, F.T., "Energy Potential through Bioconversion of Agricultural Wastes," Final report to Four Corners Regional Commission, FCRC No. 651-366-075, 7:35, 1976.

COMPOSTING TOILETS:
A VIABLE ALTERNATIVE

Presented by: David Del Porto
Prepared by: David Del Porto, Allen Field

INTRODUCTION

In this last quarter of the 20th century, we face serious problems in the areas of water supply, water and land pollution waste disposal, high costs and taxes and energy. The more serious these problems become, the more they encroach on our health as individuals and as a society. One day we hear about people being exposed to toxic substances in their drinking water, while the next we hear irate citizens damning environmentalists for high taxes and unemployment. We express concern over the impact of strip mining for coal and then hear about how many tanks, bombers and missles we are selling around the world to help balance our imports of oil. We run the risk of damaging the quality of our lives in a wide range of ways, from our health to our human relations, and even to our international relations.

While we, the citizens, hope for solutions to come along, we are often inadvertently contributing to the problems that we may be trying to solve. Some of us may busily drive around to meetings to promote public transportation, but at the same time we are not only burning up a limited resource and polluting the air, but also paying money to the oil and auto industries which in turn lobby against public transportation. While we say how awful it is that the Colorado river no longer flows to the ocean or that the rivers are so polluted, we continue to flush 30 gallons of water down the toilet daily. We cannot put all the blame onto others. It is often our behavior that is merely being serviced by others.

We cannot, then, afford to always wait for someone else to solve the problems; changes are coming too slowly to keep pace with the deterioration of our environment. For example, while the Environmental Protection Agency struggles with the reduction of the amount of water pollution from sewage with moderate success, the solutions they promote (which usually involved expansion of sewered systems and treatment facilities) put an additional strain on our supplies of water, energy and money. We can work for institutional change at all levels, and this no doubt is important. However, in the face of institutional resistance to change and its often limited vision, we can also take individual action, and only time will tell which approach will have greater effect in the long run.

By taking individual action we not only reduce our own negative impact on the environment, but we also become an influence on those around us who may well follow suit. By changing our own behavior we also educate ourselves in the process, sharpen our perceptions of an area of environmental concern and find ourselves committed to it. We are then on a firm footing to encourage others, and nobody can tell us to practice what we preach.

THE FLUSH TOILET

There is an innocent-looking item in our houses that touches on a surprising number of issue and gives us the opportunity to take individual action, namely, the flush toilet. Berkeley architect Sim Van der Ryn has imagined how future archeologists might interpret our toilet system: "By early in the 20th century, urban earthlings had devised a highly ingenious food production system whereby algae were cultivated in large centralized farmlands and piped directly into a ceramic food receptacle in each home." Surely this notion is no more absurd than our practice of mixing one part of excreta with one hundred parts of precious clean water, transporting this great volume through miles of pipes to a treatment facility where precious money and energy is expended in an unsuccessful attempt to separate the two, and releasing the effluent to our waterways where the nutrients and treatment chemicals seriously disrupt or destroy aquatic life as well as the drinking quality of the water. In this process, nutrients that came from plant life on farms end up misplaced in the waterways where they do not recycle.

Specifically, the flush toilet causes problems in the following ways: it disrupts the natural water cycle by consuming large quantities of water that are not returned to the areas they came from; it pollutes the water that is used as a vehicle to transport the excreta; it pollutes the land that is needed to dispose of sludge from septic tanks or treatment facilities; it creates large direct and hidden costs to design, construct and maintain the elaborate toilet system that extends from pumping fresh water from the ground to monitoring the effluent from a sewage plant; it requires a large amount of energy for construction and operation. These are the major areas of impact, although there are ripple-effects into many other areas, such as the medical problems for those who drink contaminated water or the rapid depletion of a farmer's topsoil.

DRY TOILETS

Of the number of alternatives to flush toilets, the one that shows the most potential for alleviating the above problems is the composting, or humus toilet. It is no coincidence that it is also the alternative most closely duplicating natural life cycles, for in the long run we must ultimately harmonize with the natural processes or become extinct. The composting toilet duplicates natural aerobic decomposition (a process that involves aerobes, which are air-dependent bacteria) by containing excreta and promoting the evaporation of excess moisture and the rapid scavenging action of aerobes to produce a small amount of fertile humus. Composting toilets can be divided into three general categories: there are those you can build yourself at a low cost; there are models that are rather large and costly but once installed do the job with little maintenance and little or no energy consumption; finally, there are those appliance-like models, such as the "Humus" that are less expensive and easier to install than the large models, but require somewhat more maintenance and energy.

Although you may run into more resistance from local and state officials with an owner-built composting toilet, a properly built one could prove to be perfectly functional and safe.

The person who builds and maintains a composting toilet should be careful about three major possibilities for contaminating the environment: the first composting chamber should be carefully constructed with a concave, waterproof bottom so that liquids will not seep out before they have had a chance to be evaporated; the toilet should be carefully built and screened to keep out insects which can transfer pathogens; the piles will need turning every week or two through an access door to keep the process from going anaerobic and to bring the cooler outsides of the pile into the middle, and the tool used for this should be kept in a way (perhaps inside each chamber) so as not to allow transfer of pathogens.

If you cannot build your own composting toilet but want a system that operates on little or no energy, you should consider one of the large models such as the Toa Throne. These models consist of a four to six foot long fiberglass or plastic composting chamber mounted at an incline. To install one of these models one must cut a hole in the floor of the bathroom-to-be and connect the toilet stool to the upper end of the container. The remainder of the container extends below the floor into the basement or crawl space. If such an arrangement is difficult, an extension could be built onto the house to accommodate it. It is also possible to connect a chute from the kitchen to the container to dispose of kitchen scraps. Given the complexity of installation, they would be especially practical as part of the plan for a new house. The excreta drop from the toilet to the upper end of the sloping floor of the chamber, and aerobic decomposition starts in the matter. Over a period of months matter slowly slides down the tank, glacier-like, as new deposits build up behind it. There are a series of vents to provide air to this sloping mass, and the vapors are vented out the vent pipe. In theory and normal practice this system isolates the old compost from the new, but if someone pours a large quantity of liquid down the toilet, it could run from the fresh deposits all the way down to the old composted matter. However, as we shall see later, this does not appear to have arisen as a problem. The waste needs to be removed from the lower end and taken to the garden only every year or two. Since the aerobic process produces its own heat and the composting chamber is in a basement that is not too cold or if the tank is insulated against the cold, the process will work well. Both these large inclined composting toilets and the home-made models can accommodate the output from large families of up to six members and extra large models are available or can be built.

The SOLAR ONE HOLER™, the first of the SOLTRAN™ series of self-sufficient sanitary systems, is now available. By combining the virtues of solar energy and composting, the SOLTRAN™ system is able to provide both an ecological and an economical alternative to sanitary requirements in remote sites, such as parks recreation areas, islands, and camps where environmental conditions demand a pollution-free facility with minimal servicing requirements. SOLTRAN™ was developed by ECOS under contract with the Appalachian Mountain Club (AMC) who, in conjunction with the US Forest Service, will be installing the units in eastern mountains and islands under their care. The objective was to deliver a sanitary system which would be in harmony with the fragile ecology of the White Mountain National Forest in New Hampshire, an area of breath taking beauty which annually attracts hundreds of thousands of hikers and campers. As traffic has increased, the AMC has taken interim measures including helicopter pick-up of human excreta in 55-gallon drums to protect the local environment. After a long and careful study by the AMC research group, it was determined that the composting of the organic refuse and human excreta was the optimum solution as it offers a natural process requiring no water, plumbing or chemicals while returning the valuable nutrients and organic matter to the soil. Use patterns and climatic changes tax the capacity of commercially available compost toilets. Therefore, the SOLTRAN^{TM} system was developed to respond to this problem. Through controlled solar heating of the composting process, the biological activity is automatically maintained at optimum efficiency. The logic to this system and its efficiency is further realized when it is understood that the use of recreational areas increases during the spring, summer and fall concurrent with the maximum availability of solar energy needed to facilitate the composting process during peak usage periods. The solar heat storage and compost chambers have adequate capacity to manage long periods of sunless days, as well as fall and winter seasons when both use and solar gain are reduced.

The third option in composting toilets requires no alteration to the house other than the installation of a vent pipe. These models are smaller than the inclined tank ones and are a single unit with the toilet stool and seat attached. These models get by with being smaller by accelerating the composting process, by keeping the compost at optimal temperatures for bacterial action with an additional heating system and by using a fan to keep a good supply of oxygen passing through the compost. The electrical input of the different models is typically about 200 watts, but the main electrical consumer is the heater which is on only intermittently. How often it goes on depends on how much heat is being generated by the bacterial activity and how warm the house is. As a general rule of thumb, one could view the consumption as a continuous 100 watts. It would, then consume about 72 kwh/month as opposed to 75 kwh/month for a "regular" refrigerator or 160 kwh/month for a frost-free refrigerator. These toilets can accommodate year-round use by up to six people, but only a few days of use by much greater numbers. Every month or so the underside of the composting mass should

Solar Architecture

be gently scraped or raked to make the dry and processed humus fall into the collecting tray. There is usually a built-in scraper for this purpose. Every few months the contents from the tray should be buried in the yard or flower garden. These toilets, like the larger inclined ones, are made from non-corrosive materials and should last many years with proper care, if not for the lifetime that some manufacturers suggest.

There is a wide variety in the costs for the various alternatives. Providing you can provide your own labor, the owner-built composting toilet could cost as little as $100 in materials. However, if you have to pay a contractor to build it, it might almost cost as much as the small, power-assisted models which range from about $700 to $950. The inclined tank models range from $1,000 to $2,000 depending on the make and size.

Although composting toilets are especially well-suited for seasonal use in rural areas where there is no ready access to quantities of water, they can also be used year-round in the urban setting. Even a person living in a city apartment could use one of the electrically-assisted models as long as a vent pipe could be installed. Systems have been designed for apartment buildings in which excreta are moved by screw-type mechanisms through pipes to large composting containers in the basement. However, much of the beauty of the simplicity of the composting method is lost through such complexities. It is relatively simple, though, to mount two toilets over the upper end of an inclined composting chamber with one (or both of them installed upstairs and connected to the chamber with a vertical chute, although in this case an electric fan would probably be needed in the vent pipe and the chute might get soiled due to its length. For city dwellers with compost toilets, it would not be necessary to have access to a yard to bury the humus produced. Although it would be wasting the value of the humus as a soil fertilizer and conditioner, it could be discarded along with other household trash since it would only be a couple of pounds per month per person of dry, safe humus. Figures 1 through 4 illustrate various types of dry toilets.

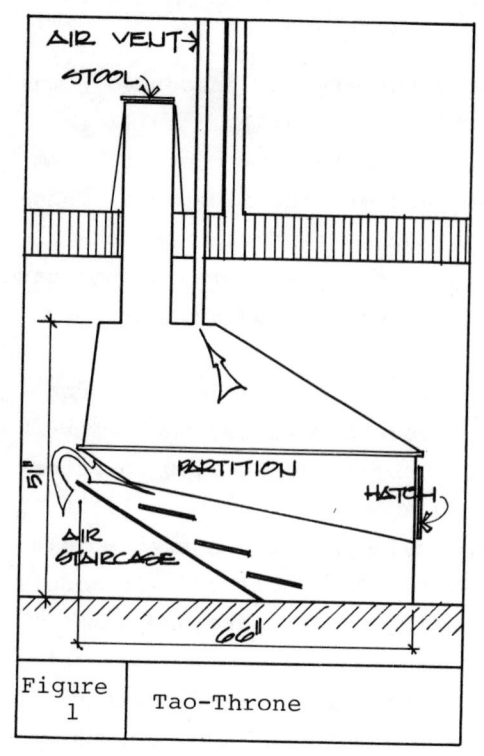

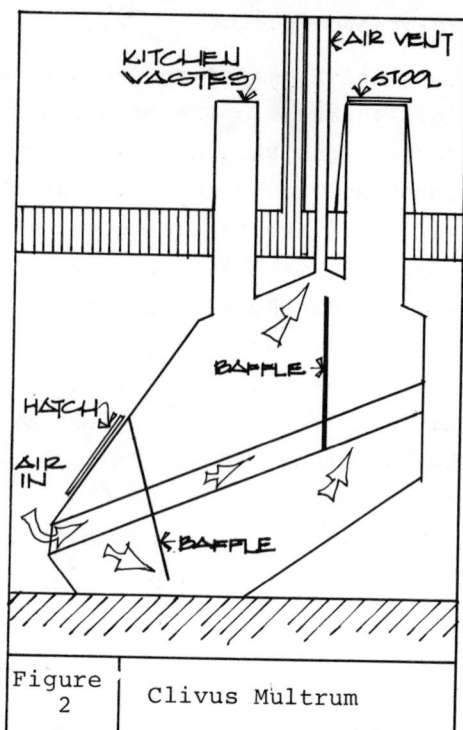

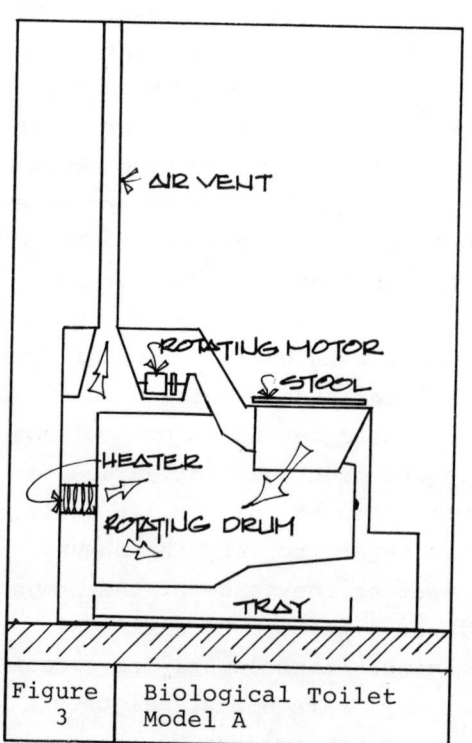

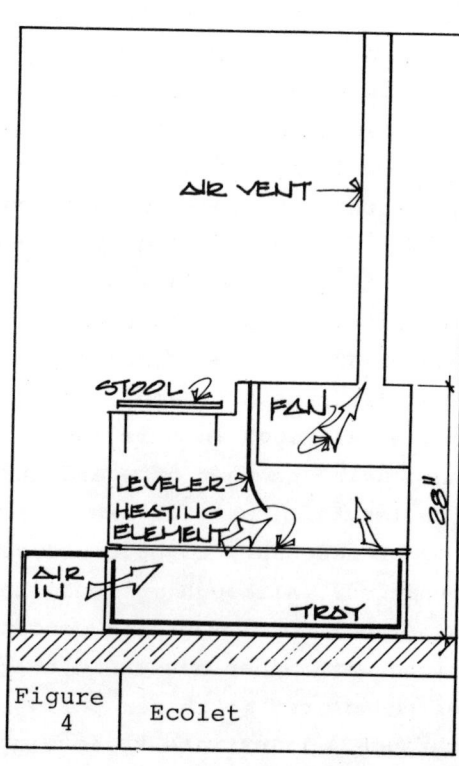

Figure 1	Tao-Throne
Figure 2	Clivus Multrum
Figure 3	Biological Toilet Model A
Figure 4	Ecolet

Solar Architecture

The following are frequently asked questions and answers regarding the dry toilets.

1. Does the toilet smell?
   No. Composting is an aerobic process which does not generate odors usually associated with outhouses. The fan mechanism ventilates the container and exhausts odors through the vent pipe.
2. Doesn't the toilet fill up?
   No. Human waste is 85% water. The remaining solids are further reduced through the biological action of composting. The warm air which is circulated through the container evaporates the liquid.
3. Will the composting process stop, if the toilet is not used for a while?
   No. The decomposition and composting processes will continue at a slower rate. The organisms which are responsible for composting create heat by their own biological activity. The odors and liquid are vented by natural convection (chimney effect) through the ventilation pipe.
4. Can you put garbage into the toilet?
   Yes. Any organic substance should be put into the toilet. This includes toilet paper, food scraps (chopped finely or shredded), sawdust, grass clippings, etc. Washwater must be disposed of separately.
5. Is the final product (humus) safe and sanitary?
   Yes. After one year of composting, humus is as free of disease-producing organisms as any good garden soil. Few pathogens can survive competition from ordinary composting organisms and the temperatures in the composting toilet, which range from 70 to 160 F.
6. How much humus accumulates yearly from the composting toilet?
   For a toilet in use full-time by one person, a total of 60-70 pounds of humus could accumulate after one year.
7. How often should accumulated humus soil be removed?
   Depending on model and frequency of use, removal periods range from 3 months to 2 years.
8. Can household cleaners be used in the composting toilet?

No chemical germicides such as bleach or bowl cleaners should be used. Should a germicide spill into the composting material, the organisms responsible for the compost process will be destroyed in the spill area. It is recommended that mild, all-purpose soaps be used to clean the outside of the toilet.

9. Will the toilet tolerate overloads?
   What if you are entertaining friends?
   Yes. If you anticipate heavier than recommended use, prepare ahead of time by shredding newspaper and adding it to the waste pile as space allows. Sawdust, peat moss or any other absorbent organic materials can be used also.

COMPARISON

An important area for comparison between conventional systems and composting toilets is direct and indirect costs. People are unlikely to change their behavior unless there is an economic incentive. Direct costs of having a flush toilet include the following: the cost of buying and installing the toilet itself and the plumbing within the house that services it; the costs of metered water bills or the installation and maintenance of an independent water pumping and delivery system; the cost of connecting to a sewer line or the installation and maintenance of a septic tank system; sewer use charges, if present; electric costs for running a water pump, if present. For people already with their own septic system or sewer hookup, savings from reduced water consumption and sewage output realized from converting to a composting toilet would pay for the conversion after a few years. This is especially true, if such people are already paying the minimum monthly rate for water. In that case about the only possible savings would be through less frequent pumping out of the septic tank, which would not likely save more than $30 per year. The reduced loads on the water supply and sewage treatment systems are significant when composting toilets are used, and it is unfortunate that at present this reduced load is not reflected in costs. (The Environmental Protection Agency calculated costs at $5/1000 gallons.)

Solar Architecture

When new housing construction is considered, however, the composting toilet is competitive in terms of costs. The cost of flush toilets, toilet-related plumbing and septic system is likely to run over $2,000 per installation. An expensive composting toilet, at $1,500 installed, and a gray-water disposal system (to be discussed later), at $500 installed, would still only come to $2,000. The cost to connect to a sewer is highly variable, but normally a few hundred dollars. Sometimes there are special one-time charges or annual assessments as well as monthly charges. A study of the proposed sewering of the town of Groveland, Massachusetts indicated a one-time charge of $3,400 (which would include the hookup and plumbing work) would be required in addition to annual tax increases and use charges totaling $285 per year for one residence. Whether it would be less expensive to install a composting toilet and gray-water system than to connect to a sewer depends on the situation, but it clearly would be in the case of Groveland.

In the area of direct costs, then, the composting toilet is not always a definite improvement, but when indirect or hidden costs are taken into account, the contrast is dramatic. The town of Pepperell, Massaschusetts, for example, which is at present using individual septic systems, is being forced to consider constructing a sewer system in order to improve water quality in the area. A study shows that the total construction costs for treatment plant, interceptor sewers and lateral sewer lines, divided by the number of homes to be served (300) would be $14,715 per home. This figure does not include the cost of in-house fixtures and plumbing, nor does it include the costs of operation and maintenance of the plant or the interest on outstanding bonds or other indebtedness, all of which is dependent on local funds. The cost of operation and maintenance may in the long run equal the cost of construction. Thus, the cost per home over a twenty to thirty-year period might be over $30,000. In contrast, due to the economies of scale, it has been estimated that a high-quality composting toilet and gray-water treatment system could be provided, installed and periodically inspected for a cost of $1,024 per home. One might think with these startling figures that people would be scrambling to get composting toilets, but the problem is that the high costs of sewering are hidden. Over 90% of the cost of plant planning, design and construction is paid out of state and, particularly, federal revenues. And, this is seen by the citizen as "high taxes". Likewise, much of the local share is hidden and perceived of as property taxes. Although in the short run the costs are shared by many taxpayers, it must be remembered that there is no special magic in the tax system, and that in the long run the typical homeowner will end up paying the enormous costs of sewering the home.

The final area of comparison is one of energy consumption. It is hard to estimate how much energy is consumed in creating the raw materials and assembling them into a finished product, such as a septic tank, a sewage treatment facility or a fiberglass composting toilet. The simplest way to view energy consumption in manufacture is through costs. If some process was highly energy intensive, it would be reflected in high costs due to the cost of the energy consumed. Thus, the previous sections on costs generally cover this area. The question of the energy consumed by the operation of excreta disposal systems is more amenable to analysis. All conventional water-borne excreta disposal systems require energy to pump water to the toilet, except for the rare situation of gravity-fed water. Septic tanks do not consume energy, although the trucks that come to pump them out do. One estimate is that a 309 million gallons per day waste treatment system, such as the one recently built for the Washington, D.C. area, will daily consume 900,000 kwh of electricity, 45,000 gallons of fuel oil and 500 tons of chemicals (which took energy to produce). Based on a family of five flushing away 150 gallons per day of sewage, this would translate into an energy consumption per family per day of 1/2 kwh electricity, 1/5 pint of fuel oil and 1/2 pound of chemicals. Comparisons are difficult, but this could be seen very roughly in the same range as the 2 kwh per day required by the most energy consumptive composting toilets. Those models using only an electric fan or no power at all would clearly use less energy

Solar Architecture

than a sewage treatment plant to process to completion the same amount of excreta. It is clear, then, that composting toilets consuming no energy in operation would only represent a one-time energy cost in manufacture and would place less demand on our energy resources than either septic or sewer systems. Whether power-assisted models would likewise require less of our energy resources, depends on what factors are taken into consideration.

LIMITATIONS
No description of composting toilets would be complete without mentioning the limitations to its use that can be discerned. At present, to convert to a composting toilet system requires an initial cost of at least $700 (unless you build your own) and unless the current laws are changed, one is unlikely to recover those costs in the short run. There is a small amount of thought that must be given to maintaining the biological composting process--it is not as simple as flushing a toilet. Of course, this might be seen as beneficial and educational rather than a limitation. There is also a small amount of work involved, the power-assisted models requiring periodic scraping and emptying every month or two. There is the problem of people's senses of aesthetics and values. Those people who are hypersensitive to the subject of human excreta and who can see it only as an embarrassing, "dirty" waste to be forever rid of might have difficulty living with the composting toilet. Those who have no awareness of or concern for environmental problems, combined with a surplus of money, would find little reason to get involved with composting toilets. Finally, some people might find the appearance of the units themselves aesthetically displeasing and not be enterprising enough to cover up the fiberglass and plastic or otherwise adapt them to their tastes.

One obstacle to the use of composting toilets is health laws. It has been shown that composting toilets provide an effective method of controlling diseases transmitted through human excreta, and it is quite evident that septic and sewage systems

have largely failed to contain fecal contamination, especially
of water, yet the health laws have been slow in changing.
Swedish and Norwegian authorities have done extensive testing
of composting toilets for health hazards and have found them
to be satisfactory in terms of sanitation. In Sweden and Norway there are over 2,000 composting toilets of just one of the
over twenty makes available. There are probably less than that
number installed in all of the United States. A number of
states, including Washington, Maine, Massachusetts and New Hampshire, have approved of composting toilets, providing there is
also an approved method available for the disposal of gray-
water. Approval from local boards of health, however, is usually still required.

Another obstacle to installing a composting toilet is plumbing
codes. The plumbing code in Maine has been rewritten to allow
the installation of such systems, but in most states it has
not and variances must be sought from the local building and
plumbing authorities.

As has been mentioned already, a separate gray-water system
is needed in conjunction with a composting toilet to dispose of
washwater and other liquids. This water is largely free from
excreta except for small amounts derived from washing soiled
clothes and diapers. The most likely hazard to the environment
from gray-water at present is phosphorus from cleaning agents.
The reduced volume of water, excreta, nutrients and BOD in gray-
water allow for a relatively small filtration system and leach
field. Some states have drawn up regulations for reduced size
septic systems to handle gray-water. Existing septic tanks
may also be used for gray-water disposal, as may existing sewer
connections. There are also filtration systems and separating
systems on the market for around $500 to handle gray-water.

CONCLUSION
In conclusion, composting toilets stand out as the method of excreta disposal with the most desirable traits and the least
potential for damaging the environment. It is the only method

that is aerobic, that does not consume water and does not have to involve energy consumption. In addition, it is comparatively safe in terms of health, simple, decentralized, inexpensive and produces a product of some value. Institutional resistance to change appears to be responsible for the continued concealment of the true attractiveness of compost toilets in terms of cost. This may be seen as selfishness on the part of public officials who do not want to "rock the boat," but it must be remembered that so many people in and out of government are simply preoccupied with their jobs and do not have the time or energy to look into alternatives. It is simply that we have a large amount of human momentum traveling in the wrong direction. However, its growing acceptance in other countries and the succession of changes in state laws taking place in this country indicates we will be hearing much more about composting toilets in the years to come.

BIOGRAPHICAL SKETCH

David Del Porto is the president of ECOS Inc. of Boston, Massachusetts. ECOS is a corporation selling a variety of dry toilets in the United States.

ENDNOTES

1. Clark, Peter, Natural Energy Workbook, Visual Purple, 1974.
2. Golueke, Clarence, Composting, Rodale Press, 1972.
3. Leich, Harold, The Sewerless Society, Bulletin of the Atomic Scientists, November, 1975.
4. ECOS, Inc., Literature on Composting Toilets, published by ECOS Inc., 12 Imrie Rd., Allston, Massachusetts.
5. Literature on Composting Toilets, printed by Clivus Multrum USA, 14A Eliot St., Cambridge, Massaschusetts.

THE ROARING FORK RESOURCE CENTER

Presented by: Gail Weinberg
Prepared by: Heidi Hoffmann

INTRODUCTION

Renewable resources will be the key question in dealing with energy in the future. A network of energy information dissemination at the grass-roots level is vital to the future of energy conservation and development of an appropriate energy technology.

The Roaring Fork Resource Center (RFRC) was established in late 1973 in response to a growing public concern of the need to use out natural resources wisely. This non-profit, educational organization is dedicated to promote, direct and develop opportunities related to energy conservation and the exploration of alternative energy sources and design considerations. The specific goals are:
1. To develop and experiment with appropropriate technology and design alternatives related to optimal consideration of natural resources.
2. To conduct ongoing alternative energy experiments.
3. To experiment with a variety of materials that contribute to the optimal design of energy systems.
4. To direct the design and construction of living units and other structures that would explore and demonstrate the optimal design considerations of energy systems.
5. To provide participatory learning environments.

6. To sponsor public educational seminars on local, regional and national levels concerning the most efficient utilization and preservation of natural resources.
7. To provide a regional library and clearinghouse that contains research data and information on related topics.

RFRC EDUCATIONAL PROGRAMS
The RFRC was founded on the premise that the integration of appropriate energy technology into communities through an educational process is vitally important. The Resource Center facilitates a variety of symposiums, workshops, college-level courses and citizen participation programs in the Roaring Fork Valley. Community educational programs are conducted throughout the year. Participants are involved not only in the classrooom experience, but also in the actual construction of prototype demonstrating energy conservation or utilizing alternative energy sources. One such program has culminated in the design and future construction of a solar-heated community greenhouse.

SOLAR GREENHOUSE FOR COMMUNITY EDUCATION
This solar-heated greenhouse employs a unique solar concept, as well as a means of food production in a severe climate, water conservation in an arid, mountainous region employing an organic waste treatment system and an essential contribution towards greenhouse horticulture. All are valuable prospects for a 9,000 degree-day climate.

The solar system for the greenhouse is a hybrid passive/active system. The greenhouse structure is integrated into the south-facing glazing and contains an array of movable sun-louvers that provide an active solar air collector, passive solar gain, glazing insulation and summer shading (Figure 1).

PUBLICATIONS
A major ongoing RFRC activity is the publication of the Sunjournal, a quarterly energy periodical that features state-of-the-art articles, technology briefs, regional activities, calendar and advertisments from companies involved with

Solar Architecture

alternative energy and each spring issue contains the abstracts for the upcoming Energy Forum. A recent project for the Resource Center was the preparation of an informational pamphlet explaining the Solar Features of the Pitkin County Air Terminal, one of the country's largest passively solar heated buildings.

LIBRARY/CLEARINGHOUSE

The RFRC maintains an office/library/clearinghouse in the Wheeler Opera House next to the Aspen Mall. The Board of Directors and staff encourage local and tourist traffic to peruse through the library, view the displays and ask questions. In addition to the many books, periodicals and reprints the Resource Center has requested and received funding for Pitkin County Library and instrumentation acquisitions.

SOLAR ARCHITECTURE

Of significance, the RFRC has been instrumental in working with the local government and public agencies in the integration of

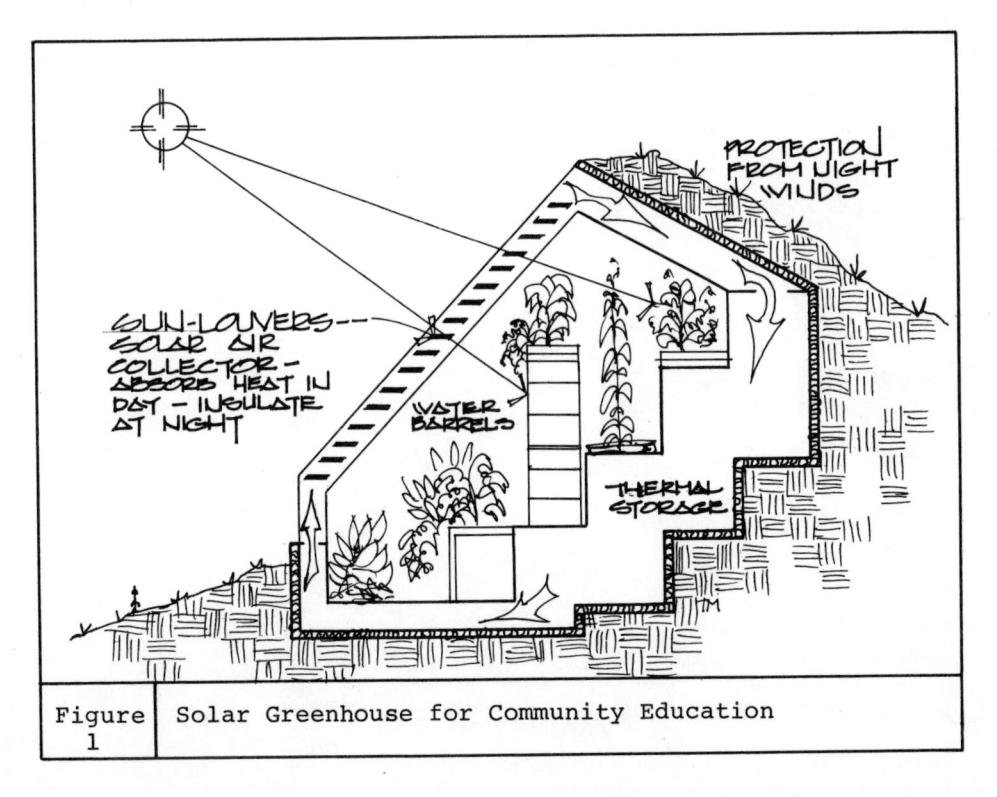

Figure 1. Solar Greenhouse for Community Education

alternative energy sources and energy conservation concepts with policy and demonstration projects. On June 5, 1975, the Pitkin County Government passed "Energy Conservation and Thermal Insulation" amendment to the Uniform Building Code. Today there are over 80 homes in the Valley attesting to community awareness of solar architecture.

The continuing operation of the RFRC is made possible by tuition from various educational programs, subscriptions to Sunjournal and sale of publications, board of contributors, Pitkin County funding of special projects and the energy from a generous corps of volunteers.

BIOGRAPHICAL SKETCH
Gail Weinberg is a land planner and Co-Director of the Roaring Fork Resource Center in Aspen, Colorado.

ASPEN COMMUNITY SCHOOL ENVIRONMENTAL LABORATORY

Presented by: Paul Rubin and John Katzenberger
Prepared by: Paul Lieblich, John Katzenberger, Paul Rubin

During the past year the environmental education program at the Aspen Community School, near Aspen, Colorado, has been expanded on many levels into the five-year-old to eight-year-old groupings. At this point, the main goal of the program is awareness of the environment as it relates to natural functioning, to man's introduction into the natural environment (as opposed to man-made environment), his use or abuse of the environment and the probable consequences for the future, given the varied directions that man can take.

To develop the awareness, children first explore the natural environment (i.e., plants, animals, earth) and its cycles. The emphasis is placed upon the interdependent nature of these cycles. Field trips and hikes are taken, experiments with plants are performed and animals and the basic earth sciences are observed and discussed. Relationships and parallels develop between the child's more personal world and the natural environment. Questions are explored along the lines of: where do we get paper, plastic, electricity, water, aluminum, tin, etc.? Once having developed this sense of use and origin of resources, the child explores life without the resource, thus establishing some reference point for the significance of the studies. This information is then built into the sense of

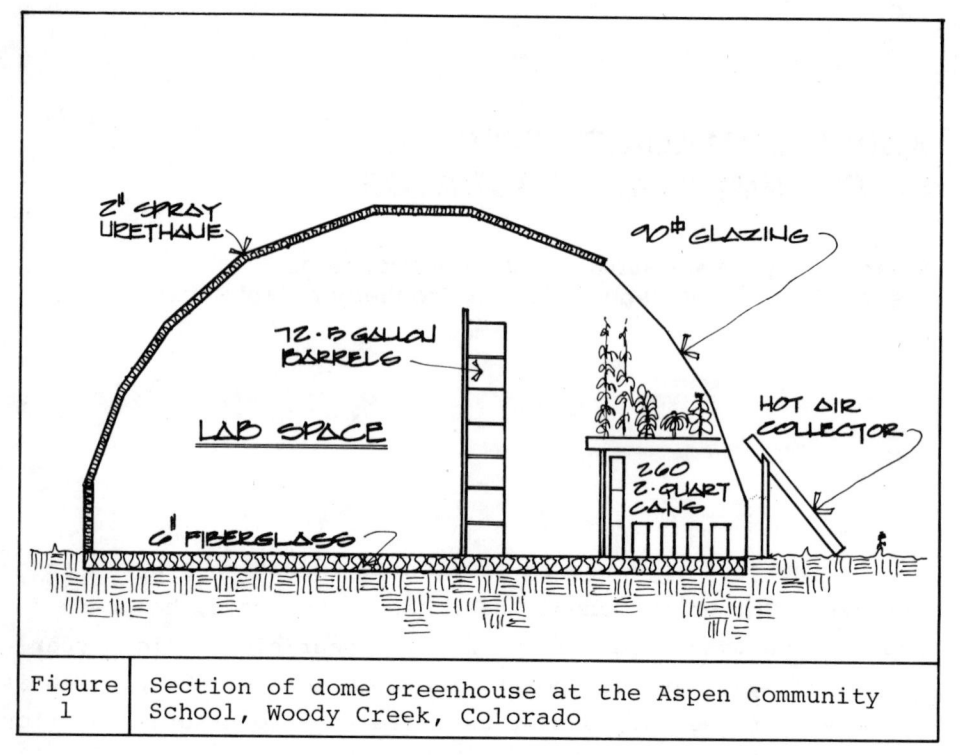

Figure 1. Section of dome greenhouse at the Aspen Community School, Woody Creek, Colorado

cycles that was developed with regard to the natural environment. In a hands-on sense, this translates to having the children establish their own recycling center for the garbage in school or making their own recycled paper and using that paper in their personal work.

With these relatively general experiences serving as background, the children launch into more specific projects and experiments which allow the children to further explore their total environment. The solar dome at the school (Figure 1) is introduced and its functioning explained through experiments that deal with reflection and absorption of light and heat, and materials that insulate store heat. With the information gained from these experiments, decisions are made on how to better insulate the school and the result is in the student-produced insulating panels for windows in the school. Follow-up record keeping on fuel use for the school, and feeling the warm school in the morning, indicate that theoretical efforts have had a practical pay off. The solar dome is further explored as the students grow plants in the dome even though outdoor temperatures would prevent such an activity.

Integration with more formal academics is apparent throughout the program. In math, the students determine weight of collected aluminum and on a per pound basis calculate the payment due from the County Recycling Center. Mass of the various materials is considered, classification of materials is explored, and cycles, graphs and time tables are made to collect and assess data. Language Arts are equally consumed. Directions for experiments are read and written. Logs of activities are kept and reflected upon.

The students are currently finishing construction of an active air collector. The test data will be gathered by the students. This collector should be able to deliver 700 Btu's per sq ft of collector. Since there is no nighttime load, it will require no auxiliary heating system. Improvements in insulation in the interior will further increase the dome's efficiency.

We are currently engaged in expanding the program through a grant entitled "Project Shelter." This program will further develop multi-disciplinary approaches to education and will delve into new and exciting environmental education possibilities through the completion of a large environmental learning center on the campus.

It is hoped that with the environmental education program at the Aspen Community School, the children will become more aware and develop tools with which they can explore their total environment and view it "holistically." Finally, it is hoped that the energy and environmental decisions for their future will be the right ones for all.

BIOGRAPHICAL SKETCH
Paul Rubin and John Katzenberger are instructors at the Aspen Community School near Aspen, Colorado. Paul teaches environmental education and John teaches science and math.

WILDWOOD SCHOOL: A PLACE TO EXPLORE THE ENVIRONMENTAL ETHIC

Presented and prepared by: Bob Lewis

With a new and constantly growing concern for the protection and improvement of our environment, it is both timely and relevant to foster an awareness of nature and of our place in nature at the earliest educational level; namely, the Pre-School. In these formative years, when many of the most significant concepts are formed and values are shaped, an Environmental Ethic can most easily germinate in the child's consciousness. It is the objective of the Wildwood School to help develop this Environmental Ethic: an understanding of the ecological balance of this planet and the mutual interdependence of man and nature.

Most planners and ecologists agree that by the year 2000 it shall be known whether an acceptable quality of life can be maintained on earth. The significance of this situation has relevance for all citizens and it has most relevance for

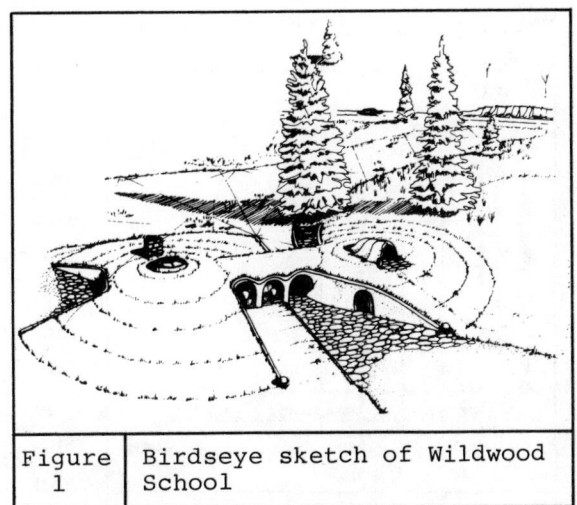

| Figure 1 | Birdseye sketch of Wildwood School |

the young. Given a hypothetical life span of eighty years, it
is easy to see from Figure 2 that pre-school children can look
forward to spending 5/8ths of their lives beyond the year 2000.
Therefore it behooves adults to make every effort to provide
opportunities in environmental education for the young who
have so much at stake. Wildwood's program is geared to provide exciting learning experiences that will lead to an <u>Envir-
onmental</u> <u>Ethic</u> at an early age. Only by adopting an <u>Environ-
mental</u> <u>Ethic</u>, in the future throughout all strata of society,
can this country avoid the strife that surely follows when
demands exceed resources, and pollution levels of land, water
and air rise beyond tolerable limits. America can no longer
continue to grow in population and create greater and greater
demands on energy resources and other non-renewable earth
treasures. New ways must be found to think small, use less,
and re-cycle everything. Our greatest resource, the inexhaus-
itble sun, can lead us to an enriched new way of thinking and
living. In the words of the "Friend of the Earth", David
Bower, "we can no longer steal from our children". He was
speaking of our greedy use of their energy and resources, that
they will need for their own lives, and the lives of children
for generations to come.

Children have first hand opportunities to observe Nature since Wildwood is in an unspoiled mountain forest. They have direct involvement with plants and animals in the field as well as in the learning modules themselves. Working and play-ing with nature will lead to an awareness of its complex interrelationships which is the basis for understanding the changes in the physical and biotic environment which have been introduced by man. This understanding in turn is

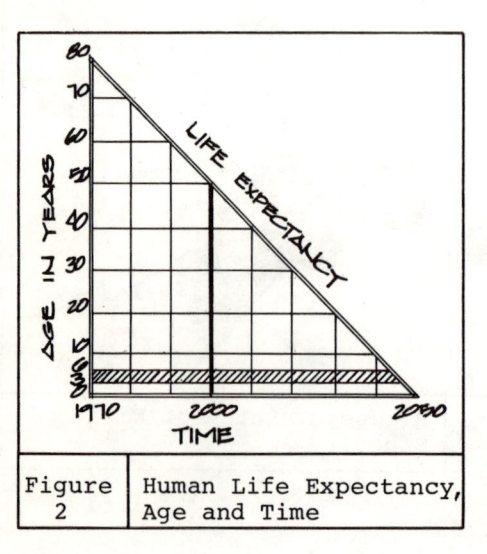

| Figure 2 | Human Life Expectancy, Age and Time |

Solar Architecture

the indispensible prerequisite for Environmental Consciousness.

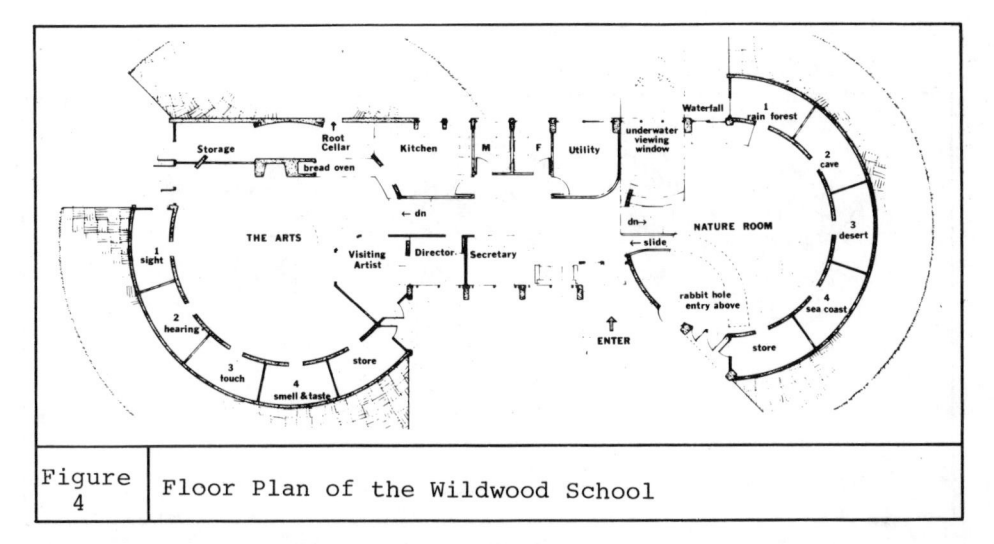

Figure 3 — Cutaway of the Wildwood School

Figure 4 — Floor Plan of the Wildwood School

Wildwood's Contribution to Society

1. Wildwood has pioneered new and inexpensive school construction techniques and serves as a demonstration center for those interested in energy saving techniques.
2. Wildwood demonstrates that a construction crew, composed primarily of college-age men and women, can make important contributions which will benefit themselves and, at the same time, lower construction costs.
3. Wildwood emphasizes a new approach to pre-school education which utilizes the natural environment as the learning environment.

4. In the future Wildwood will serve as a research and observation center for school designers, child behaviorists, ecologists, artists and all others interested in how young children interact with the environment.
5. Finally, Wildwood could be the long overdue catalyst to begin the chain reaction of similar efforts throughout the country which will require a much-need <u>Environmental Ethic</u> in future generations.

The Creative Process

Traditionally, schools expect input and output to be in the same medium. However, we encourage children to learn (input) about Natural History and will encourage their activities (output) to be in another medium--in the broad field of Art. This we feel is the essence of the creative process.

Figure 5. Nature/Art. Input/Output.

Nature and Art

As we see it, our duty as educators is not to "teach children" but to encourage individual discovery and interpretation based on a wide range of experiences. Children love to learn about and play with plants and animals, sand and leaves, stones and water. As they play, they learn. Equally natural, it seems, is the attempt of young children to express and translate their experiences through drawing, painting, clay, singing, dramatic play, music and storytelling. Art is the medium through which the young child, not yet dependent on verbal skills, most easily and naturally expresses his ideas and emotions. Thus

Nature and Art become the twin foci of the Wildwood program, constantly influencing and reinforcing the child's growing awareness and responses to his or her environment.

BIOGRAPHICAL SKETCH

Bob Lewis is the founder and director of the Wildwood School in Aspen, Colorado. Mr. Lewis is very active with environmental and energy activities in Colorado. Mr. Lewis is also a member of the Board of Directors of the Roaring Fork Resource Center in Aspen, Colorado.

MEANING, ECOLOGY, DESIGN, ETHICS

Presented and prepared by: Wallace W. Wells

INTRODUCTION

In 1972, at Stephens College, the Beta Project was begun. A group of students and faculty agreed upon one priciple: "Ours is a civilization in crisis." Most of the members of Beta Project hold that the crisis is culturally pervasive. Therefore, it is quite difficult to bring into perspective. However, this principle is the basis for continuing interaction by Beta Project members and others.

The Beta Project has both advantages and disadvantages in being located in the Midwest at a small undergraduate institution for women. It is an advantage by not having a highly specialized curriculum in technology. This leads to inquiry in technology which demands a re-evaluation of basic assumptions and the integration of technology into a "holistic" perspective. A disadvantage is that many within the academic community, as well as without, are uncertain as to the value of pursuing our unique brand of inquiry. It is recognized that while Missouri reflects a low key and simplistic pattern, it lends itself well for inquiry into the design of integrated and holistic living patterns. The raison d'etre of the Beta Project was based upon the awareness that most institutions of higher learning, by their very nature, turn out highly specialized and competitive individuals. These individuals, particularly in America, generally become affluent super-consumers. The complexity of American society makes it quite difficult for all to understand and accept responsibility for the consequences of our affluence, consumption and choices. The use of many things which use irreplaceable resources and which may be having irreversible consequences in the destruction of present life is desired. Because of complexity, ignorance and our learned desire, individuals have a blind faith that someone will save us from ourselves, while the pursuit of life styles is continued that are intellectually and humanly irresponsible as far as survival is concerned. In many institutions of "higher learning", careerism has become the order of the day, transforming education into training, principled conduct into occupational adjustment and almost completely abandoning critical pluralistic inquiry.

BETA PROJECT OBJECTIVES

The Beta Project has a set of "ideas" which are held up as measurements against which the individuals must measure their own effort. It is generally held in the Beta Project that many acceptable and encouraged practices do violence to oneself, others and the environment.

Beta Card #11; Spring, 1973 "To associate freedom with the belief that one has the right to acquire whatever thay can "afford" -- To associate freedom with the belief that standard of living has a one-to-one correlation with increased consumption -- To hold that freedom means one can have as many children as one may wish -- To associate freedom with the practices of the clever, competitive and greedy who continue to extract from the common store whatever they wish, in the name of free enterprise -- is to bring disaster upon all the earth and humankind."

Beta Project "Ideas"
1. One should integrate feeling, thinking and doing into an ethical life pattern.
2. One should always opt for low technology over high technology. Why? There are three main reasons. One, it appears

Solar Architecture

to be more rational to have life simple rather than complex. Two, for responsible conduct, one should be able to construct, maintain and comprehend the consequences of one's actions. Three, there is serious doubt that division of labor, specialization and complexity are giving us the civilization we always held they would.

3. One should recognize that interdependence is of human value, but that being dependent upon a complex and fragile network of human action and services with which one has very little comprehension, association or control is inhuman and irrational.

4. One must recognize that in most cases, one must resist a most attractive and seductive culture, if one is to create a comprehensive and responsible life pattern. While the knowledge found in our institutions of higher learning is great and glorious, most students opt for specialized training in which they focus upon the development of marketable skills rather than becoming educated. We challenge serious students to create their own careers, rather than fitting in and adjusting to ready-made patterns.

As a stimulus to learning, the Beta house was designed. Over time, it has become a way of doing Beta and making Beta inquiry. The Beta house stresses that one cannot have an external concern for shelter, food, energy and the environment and at the same time, turn one's living pattern into a responsible way of life. We must stop delegating those activities that are basic to tuning us up for living responsible life patterns. The design of this home attempts to reflect and encourage a way of life. The Beta home is a place where the individual can be challenged creatively, intellectually and emotionally and can discover new meanings of beauty and fulfillment while re-establishing harmony with others and the environment.

THE BETA DESIGN

The Beta house is an underground dwelling, 45 feet x 60 feet, with 8-inch concrete walls that are reinforced with steel rods. The walls will be insulated with 1-inch thick urethane foam from the footing up to 6 feet, then 2 inches thick up the rest of the wall and over the roof. The house will have approximately 2200 square feet usable floor space, not counting the roof garden and the two atrium gardens. For natural light and for thermal energy collection, two walk-in atriums are included; one directly facing north and one directly facing south.

The ceiling will be 6 inch reinforced concrete, with slight slope north and south from the center. The ceiling will be covered by 2 inch urethane foam insulation and 16 inch mulch and topsoil. The ceiling will handle a load of 178 pounds per square foot. A vegetable garden is on the roof. It is very interesting to live under the place where one grows his food.

The house and site area will have an extensive drainage system. All water used will be collected by the design system and recycled. The house inside will have two water systems: one potable and the other for utility use. The grey water system will be sand filtered and recycled. Cistern water will be put through a solar still and made drinkable.

The dwelling will have two self-contained waterless kitchen and human waste disposal systems (modified Clivus Multrum). For those who are myopic about the energy problem -- point of information: Next to population, the world's number one problem is potable water. Half of the water used by the average American family goes for flushing the "John".

The Beta house will be completely energy free, so far as heating and cooling are concerned. Our technological expectations are for a design temperature of plus or minus 3 degrees of 70 F, year-round. The underground design practically eliminates radiation and convection heat loss factors. Our conduction heat loss is controlled primarily by a Delta T of 10 to 15 F. For subterranean walls without insulation, it is general practice to figure conductive loss at 2 to 4 $Btuh/Ft^2$, using the simple q=U*A. Where: floor U* = 1 and subterranean walls U* = 2 in Columbia, Missouri with ground water temperature at 58 F.

Solar Architecture

The major factors in comfort are circulation, ventilation, humidity and temperature. The Beta house has fairly constant temperature and a ventilation and circulation system has been designed which utilizes unique ambient air intake controls. These controls allow for the dehumidifying, preheating or pre-cooling of air as it is brought into the dwelling.

CONCLUSION

The Beta house is a meaningful integrated design. It brings together harmoniously the psychology, economics, ideology and ethics of living in this modern ecological era. In summary, the Beta house has these features:

1. The Beta house and site are an integrated design.
2. The design is a heating, air conditioning and humidifying unit.
3. The design is a food production unit.
4. The design is a water purifiying and recycling system.
5. The design is a water-saving, self-contained, sewage disposal unit, which allows for returning nutrients to the soil from whence they came.
6. The design allows for the development of alternative electrical production and usage reduction.
7. The design commits those who live and work within it to become more sensitive to their responsibility to themselves, others and the environment.

"When the sun rises, I go to work, when the sun goes down, I take my rest, I dig the well from which I drink, I farm the soil that yields my food, I share creation, kings can do no more." Ancient Chinese Proverb.

BIOGRAPHICAL SKETCH

Wallace W. Wells is the co-founder of the Beta Project and coordinator of interdisciplinary studies at Stephens College in Columbia, Missouris. Mr. Wells is also a member of the Board of Advisors of the Roaring Fork Resource Center, Aspen, Colorado.

THE FARALLONES INSTITUTE—
EXPERIMENT IN APPROPRIATE TECHNOLOGY

Presented and prepared by: Alison Dykstra

The Farallones Institute is a small, independent association of scientists, designers, and technicians who are carrying out one of the country's first research and education programs in what is being called Appropriate Technology - the development of integrated, small-scale, diversified systems which are not wholly dependent upon finite natural resources, and centralized bureaucratic processes. As people become aware of changing social values and the realities of diminishing resources, technologies must be developed which provide a high quality of life while at the same time minimizing waste and pollution. The Farallones Institute is investigating strategies and technologies of change through research, development, and evaluation, and linking this with public education programs and training workshops.

The Farallones Institute was founded in 1974 and currently has two fully operating centers: The Integral Urban House in Berkeley, and the Rural Center in Occidental, California, which are involved in integrated research and development in the areas of energy systems, water and waste recycling (composting toilets and greywater reuse), food production, land management and habitat design. A farm center is in its early developmental stages and will be doing investigative work into low water irrigation for commercial crops, and a fourth center, concerned primarily with solar design, retrofit and technical training is due to begin operation later this year. The Institute believes that the future quality of our lives depends upon evolving a society that scales down human wants; a society in a balanced relationship with nature. Our centers are developing the tools necessary to provide options for ourselves and others to make the transition to such a society.

THE RURAL CENTER
The home location and administrative center for the Institute is our Rural Center, located on a largely undeveloped 80-acre ranch in Occidental, 70 miles north of San Fransicso. This center serves as a large-scale demonstration model and "living laboratory" for our research programs as well as our residential educational programs in Whole Life Systems and skills training. Eight staff members and 12-15 students work,learn and share in the effort to create a model for small-scale, economically self-sustaining systems of life support. The Rural Center is concerned not only with investigating systems and approaches but in training people to compete competently and successfully in the job marketplace. Our research activities provide the framework within which students can gain in-depth conceptual and technical skills in energy systems, waste recycling, land management and food production.

A Comparison Study of Three Solar Space Heating Systems
The Farallones Institute Rural Center is conducting a solar space-heating comparison study to monitor and evaluate the actual performance of three solar integrated dwellings, each with a different collecting system. These systems are examples of solar systems which are an important and viable alternative for low-cost housing in California. The study will provide empirical data on these three low-cost models which are identical in heat loss, storage and climate conditions. The controlled elements of the experiment will produce valuable insights into the cost effectiveness and comparative qualities, as well as individual functioning of these systems. In

Solar Architecture

addition, theoretical calculations can be checked against actual performance. All information and conclusions will be compiled into a handbook for designers and builders outlining the applicability of the systems, as well as the procedures for employing them.

A lack of good comparative data under controlled experimental conditions on the performance of low-cost, low-tech integrated systems will make the results of this study invaluable on a theoretical as well as practical level. By monitoring the three units qualitatively (maintenance, glare and comfort parameters) as well as quantitatively, the conclusions will start to provide a basis for preference in applied situations as well as an understanding of the thermodynamics involved. The cost analysis will form another dimension of consideration and comparison. The documentation will provide a valuable tool for the understanding, propagation and utilization of these systems, including an analysis of the range of applicability for California housing.

The three dwellings of 300 square feet each are built with standard construction techniques and State code insulating specifications. The heat loss factor for all three is 5,000 Btu/degree day and the storage capacity is 8,000 Btu/F or

Figure 1. Staff cabins, part of solar comparison study, were sited for maximum solar efficiency

about two winter days. The designs are "integrated" in that they use the building itself, to a greater or lesser degree, as collector and storage container. They represent a low-cost, low-tech simple alternative to so-called "active" systems of solar collection, storage and heat exchange.

The first system uses simple south glass as a collector and a mass of three-inch rock (approximate size) in the crawl space for storage. By using the existing glass for collection and the foundation for containment of the rock storage mass, the system is limited to the cost of a low volume fan, ten yards of rock, insulation and thermostats. The fan/rock storage system is designed to reduce the overheating associated with more passive systems. Summer cooling is achieved by blowing cold night air into the rocks. Redistribution of stored heat or cooling will be mostly radiant and naturally convected. In addition, night insulating shutters are used to equalize the heat loss with the other units.

The second system utilizes a roof pond collector and is a more complex but more controllable system than the south glass. It employs a mass of water on the roof which is simultaneously a collector and a storage medium. In intercepting the solar radiation before it reaches the living space it reduces the chances of over-heating. The heat is transferred to the space by simple radiation and is in tandom with a smaller south glass wall and rock storage. An additional feature is that summer cooling can be obtained by cooling the water at night through radiation to the night sky.

The third system employs a flat-plate hot air collector and a rock storage. This type of collector is in widespread use and is a part of the comparison study in order to judge the relative merits of the more integrated systems as opposed to a more conventional collector. It also is in tandem with a small south glass collector and achieves summer cooling by night air.

The three systems were designed to have similar heat collecting

Solar Architecture

capacities, all providing from 90-95% of the space heating needs of the dwellings. They are located with equal exposure to the sun and other climatic forces.

In addition to the three cabins in the comparison study, a fourth dwelling, which represents a typical residential application of one of the systems, was built. This cabin is an example of how a basic system of south glass and remote storage can achieve multiple functions. The south glass in this case is an attached greenhouse which simultaneously collects heat for storage and direct space heat, produces vegetables, is a site for recyclable compost from the attached Clivus Multrum and insulates the south wall from infiltration and convective heat loss. By separating the collection area from the living area, the temperature can rise above the maximum comfort level of 80F in the middle of the day and therefore produce a higher quality of heat for storage. As in the smaller cabins, the heat is directly retrieved from storage through the slab floor acting as a radiator and the option of summer cooling is present.

The cabins are complete and heating systems are functioning. During January they maintained minimal temperatures of 55 F through frost nights (approximately 25-30 F) and have performed as expected. Through a grant from Pacific Gas & Electric Company for data collection equipment, we will precisely monitor the climate, insolation and performance of the cabins for a two-year period beginning February 1, 1977. From the information gathered, a complete quantitative and qualitative analysis of the systems will be made. This analysis will provide the basis of further cost optimization of the systems, as well as verified design techniques including calculations and construction details. The material will be compiled into a designers and users handbook which will provide verified models of how the systems operate and the technical information necessary to design and employ the systems. The goal of such a publication will be to propagate the use of solar heating in residences: first, by offering dependable proof of the cost effectiveness and success of solar space heating for the assurance of banks and financial backing; and second, by offering the verified design tools of calculation instructions and construction details for builders to bring the systems into use. In addition, the data will be used for the verification of computer programs which model and simulate the functioning of solar systems and will be important tools in the employment of solar energy on a large scale.

The solar cabins are an example of a whole-life system research and development program. They simultaneously provide housing, an educational experience for students, demonstration for visitors and workshops, an ongoing research tool, and finally, published information which will aid in the acceptance and employment of solar space heating on a large scale. In this way, the broader Institute goals of developing and propagating soft technology for society at large are fulfilled while the needs of the Rural Center are satisfied.

Solar Greenhouse

The basic concepts used to achieve climate control in a greenhouse (Figure 2) are the same as those employed for solar space or water heating. First, reduce the heat loss of the structure as much as possible. This means insulating all surfaces not oriented to receive sunlight, such as the north wall, part of the east and west wall, and part of the roof. Also, reducing the air leaks common to most houses would have a major impact. Second, provide a storage medium to hold the heat for nights and cloudy days, which in our case is water-filled drums.

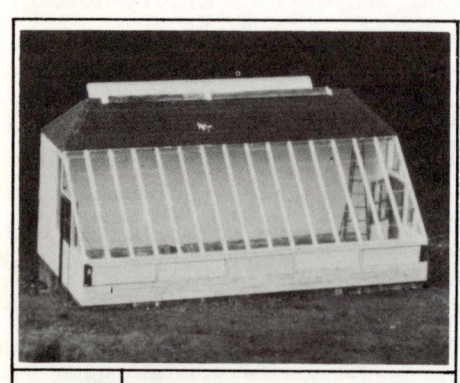

Figure 2	Solar heated greenhouse

In most green houses, the thermal efficiency to be gained by insulating the north wall has been sacrificed for lower,

Solar Architecture

standardized construction costs at the expense of massive long term energy consumption. There is, however, the argument that plants need light from all sides and are more dependent on the duration of sunlight, not the intensity. This is an important question which we plan to study by examining the comparative effects of reflective versus dark surfaces on the north walls and the effects of the absence or reduction of light from the east and west. The determination of the impact of such variables will provide design criteria for the amount of east and west walls to be glazed, the degree to which the north wall can be a dark radiant storage collector, and the degree to which the heat loss (i.e., heat load) can be minimized.

The solar greenhouse is a demonstration of the use of appropriate technology, and an example of ecological design. It serves as a low-energy-cost food production system in difficult climates, as a source of heat for an attached building, as an enclosure for biological sewage treatment, and finally, as an integrator of all of the above functions into a whole life system.

Solar Hot Water Heating Systems

The Rural Center currently has two solar hot-water heating systems in operation: a breadbox type hot-water heater and a flat-plate water heater. The breadbox heater is a purely passive, thermosiphoning system which utilizes a 90-gallon water heater, painted black and set in an insulated plywood box. The top and front of the box open to expose a reflecting surface which helps direct solar energy onto the tank, heating it up to 130 F. The shower loses approximately 15 F of heat over the course of a cloudy day and utilizes a wood-burning stove as backup. The second system, a flat-plate collector, is located on the roof of the workshop at an angle of 27 degrees and heats 30 gallons of water up to 140 F. The collector itself is 15 square feet, with an absorbed plate of 3/4" galvanized water pipe soldered to corrugated metal roofing and screwed into the collector box which is standard 2x6 construction. The box is insulated with 3" fiberglass (R-7) and single glazed with 7/32"

314   Solar Architecture

tempered glass.

Wind Systems

The Rural Center currently has one home-made, sailwing wind machine in operation which was designed and constructed by students last summer. Constructed of sheet metal and plumbing and auto parts, it was initially designed to aerate a greywater oxidation pond. Upcoming plans for doing further investigation into the uses of wind will include constructing a small generating machine on our barn, setting up a commercially available Jacobs and putting together a high-speed prop for pumping water.

Composting Privy

The Farallones Institute Rural Center is monitoring the operation of two composting privies in cooperation with the Samona County Health Department and Sonoma State College. The County Health Department was at first reluctant to allow us to carry on our investigations; however, after satisfying them of our qualifications and the legitimacy of our work, permits were issued with the requirement that an independent evaluator be obtained. Dr. Ruth Blitz, a microbiologist on the staff at Sonoma State College, agreed to participate and has been very helpful to us. Graduate students in microbiology are interested in the project and eventually may perform some of the necessary lab work.

| Figure 3 | Farallones composting privy |

One aspect of the monitoring is visual inspection of the privy once a week when it is opened for turning. The presence of any vectors is noted and their path of entry sought and corrected. The accumulation of fresh material is noted for determination of loading rates; the nose test is used to detect odors and any anaerobic activity to be corrected; and

Solar Architecture

the general rate of decomposition is observed and noted.

A very important aspect of the monitoring is the temperature data. Daily readings are made with a long probe thermometer of each pile in storage for the first six weeks and a log kept of these readings. Ambient air temperatures were recorded at first but when sufficient data indicated no correlation between ambient and compost temperatures the measurements were discontinued. Pathogenic organisms have known thermal death points which are exceeded by hot composting temperatures. There is some question whether privy compost heats up sufficiently and this is the major reason for temperature monitoring.

Approximately 16 ounces of a sawdust/leaf mixture is added to the pile with each privy use. Piles are accumulated for approximately two months, then transferred to storage bins for a minimum of six months of composting. From the weekly observation it is apparent that not much composting goes on during the accumulation period in the privy. The feces/sawdust/leaf mixture does not compost well on its own, therefore the major composting for treatment purposes must occur in storage. A properly proportioned compost pile (C/N ratio about 30 or 35 to 1) should be built in storage by adding additional materials. The nitrogen-rich feces/sawdust/leaf mixture from the privy, should comprise a maximum of two-thirds the final volume of the pile. The remainder should be readily decomposable carbonaceous organic material like household garbage, straw, spoiled hay, garden wastes and grass clippings, but not additional sawdust, tree trimmings, or other woody material. At this point the user's skill as a garden composter must come into play.

Since May 1976 we have produced four batches of compost from our two-hole privy (used by an average of 12 persons per day). The first two batches were composted carefully in storage with additional organic matter, such as wheat straw and garbage, incorporated in layers with the privy mixture of feces, sawdust and leaves. These composted satisfactorily and heated up quite well. Each pile was mixed and aerated after two weeks and again heated up. The second two piles were simply transferred into storage with little additional carbonaceous material added. These piles did not heat up satisfactorily - attaining only 42-43°C and dropping down to 32-30°C after about two weeks.

A second privy was designed and constructed in the summer of 1976. It is an example of a retrofit; that is, building a composting toilet into an existing structure. The space available involved limitations, but the design demonstrates the flexibility of the composting toilet in application to existing buildings.

The basic construction of the substructure is the same, cinder block walls on a dished out concrete slab with removable plywood access doors. Additional storage and composting compartments are incorporated and material can be turned from one bin to another internally so the compost remains within the structure for a full six months after leaving the collection bin. This is a great improvement over the last privy. The restroom itself has been nicely finished in sheetrock and wood paneling. The walls and ceiling are insulated and screen windows provide adequate light and room ventilation. Hand washing facilities are provided as is storage space for toilet paper and the high carbon mixture that is added after each privy use.

The Garden
The one-acre garden at the Rural Center serves as a training and learning center for our interns and has managed to provide as much of our year-round food needs as possible. Additionally, it serves as a source of culinary and medicinal herbs, flowers and naturally, a center for inspiration and enjoyment of nature. The agricultural procedure could best be described as mixed biologically stable farming. Raised beds and intensive planting systems are used. Our planting is done in accord with the principles of humus building, companion planting, catch-cropping and crop rotation. This process in turn maximizes the bird, insect, reptile and microbiological life in and around

Solar Architecture

the cultivated area.

Our gardening and farming is based on the principle of "living soil". This means that through the use of composting and cover-cropping a high level of fertility is maintained and soil life is kept flourishing. From this perspective we pay as much attention to the process of death and decay in plants as we do in their growth. In composting we use a modification of the Indore method evolved by British soil scientist Sir Albert Howard. Piles are turned twice and usually reach a temperature of 160F. Our finished product is ready for garden use within 10-12 weeks. This year we composted close to 50 yards of animal manures (equivalent to approximately 25 loads in a 3/4 ton pickup) hauled in from various local dairy and poultry operations, as well as at least that much by volume of our own weeds, wheat straw and garden/kitchen refuse.

Our water supply comes from two small springs which provide three gallons a minute at best and usually diminish significantly during peak demand times in late summer. We have created approximately 90,000 gallons of water storage capacity, since our beginning and plan to continue water storage development. Ground water is notoriously absent in this area, making drilling a high-risk expenditure. Due to these constraints, agricultural activities have thus far been concentrated on a relatively flat and rich area of about two acres.

Figure 4 The garden is a training and learning center.

The garden area has been deer fenced and water is gravity fed from the spring source.

Due to the severe drought conditions, we have been experimenting with a variety of approaches to watering our raised beds, including drip irrigation, heavy mulching, overhead watering and hand spraying. Our experience seemed to indicate that short frequent waterings were more efficient than long, deep, occasional waterings. We plan to experiment along these lines next year and develop information for limited water irrigation in an intensively planted area.

As with the rest of the Farallones experience, the education was in the doing. Lending a hand and playing a part in the planning, care and maintenance of the garden consitute the body of learning and students with a special interest in the garden assumed major responsibilities and became an integral part of its function.

Water Reclamation
Initial studies with greywater reclamation at the Rural Center have concentrated on the development of a cost-effective, small-scale waste-water management system designed to reuse greywater for agricultural purposes. (Greywater is defined as household waste water from sinks, laundry, shower and bath; it does not include discharge from flushing a toilet). The problems of distribution, irrigation and filtration have been the focus of our efforts. Future studies dealing with the biological and chemical makeup of greywater and the effects on soil and plant growth are being developed.

A number of different distribution systems have been investigated. Our first attempt, distribution by hand from 50-gallon collection drums in the garden, proved aesthetically unpleasant, and after a month or so accumulated grease and settled food particles in the drums went septic (anaerobic). Our next system was sub-surface and utilized 3/4" black plastic tubing under a heavy mulch layer on top of an intensively cultivated bed. Grease and suspended solids clogged the tubes and they had to be flushed out periodically with hot water - a time consuming and wasteful procedure. We next substituted a larger diameter redwood culvert constructed of scrap boards in a gravel lined trench 6" wide and 10" deep and covered with sod alongside an asparagas bed. Because of the minimal velocity of the greywater due to low head, distribution was not very even along the bed.

In the summer of 1976 a settling tank was added to remove most suspended and floating solids. To utilize this less concentrated flow, an irrigation line of 1" perforated tubing surrounded by pea gravel in a 4"x4" trench was installed along a permanent forage crop bed. If this type of distribution system proves successful, it will be used more extensively for water demanding crops such as comfrey and alfalfa for forage, perennial vegetables like asparagas and artichokes, and bush fruits like raspberries and blueberries.

Another irrigation technique was tried for intensive beds planted to annual vegetables such as squash and beans. Rigid 3/4" PVC pipe perforated with 1/8" holes at 6" spacing was placed down the center of the bed under a heavy mulch layer with the holes pointing up. A 50-gallon drum set up on a platform 18" above the bed was connected to the PVC tube by a flexible hose using hand-tightened hose connections for easy coupling and disconnection. A valve on the drum controlled the flow into the line. For testing purposes we used pond water but the system is applicable to clarified filtered greywater and will be investigated more extensively this summer. We feel the idea has promise and simulates greywater distribution under adequate pressure head.

The first week of winter rains made it apparent that greywater reuse would be limited to the dry summer months. Some alternate disposal system would be necessary during the rainy season. The simplest and most conventional water disposal system is the leach line that is used with standard septic tanks.

Several different filtering systems have also been investigated including a grease trap at the kitchen drain, but sediment accumulated in the bottom and went septic causing odors. A coarse organic filter was devised to strain out food and scum utilizing various filter mediums: wood shavings, chopped straw and leaves. The organic mat was placed in a wire cage within a 50-gallon drum and could be removed and changed. This system proved to be undersized for the flow volume and amounts of filterable particles. This system might, however, be suitable for small scale residential application.

We decided therefore on a large permanent settling basin that would retain the greywater in a quiescent state long enough to settle out the majority of the suspended solids and trap all the floating solids and scum. We calculated the cost of materials for the size tank we wanted and discovered a commercial septic tank of 1500 gallon capacity was cheaper than we could build ourselves. It was purchased and installed in a day. A larger grease trap was added ahead of the tank following plans from United Stands Privy and Greywater booklet. The effluent from the septic tank now goes to the disposal field via the switching manifold.

We had avoided slow sand filters from fear of frequent maintenance problems. Again, it was a question of scale. We had been thinking 50-gallon drum units, but what was needed would have to be much larger. Our greywater consultant came up with a sand filter he designed and has successfully implemented. It is quite large (10' x 20') and quite costly, over $1000 in materials. We are planning an adaptation of the design for installation this summer. We anticipate that our design will give us a clarified effluent that we can use more extensively in the garden without the elaborate gravel-encased irrigation lines described previously.

Animal Husbandry
Animal husbandry in the "whole-life systems" symbiosis is integrated from the barnyard, to the kitchen, to the garden. At

Solar Architecture

the Rural Center, one dairy cow and two beef calves graze about 65 of the 80 acres utilizing the natural forage. In the process they reduce the potential hazard of dry grass, and at the same time return fertilizer to the soil through their waste and spread the seeds of forage plants (which cannot be digested well by ruminants). Manure from the barn and barnyard, including that of cows, chickens and rabbits, is used directly and in compost for garden and orchard. Alfalfa, comfrey, other herbs and weeds from the garden that benefit from the manure fertilizers in turn help feed all the animals. In the kitchen, milk, eggs and meat provide the Rural Center with the highest quality of food. Leftover vegetable trimmings, meat and milk by-products are fed to the animals or composted...the cycle repeats itself continuously.

Practical involvement (such as milking; feeding and watering; cleaning the chicken coop; collecting the eggs; turning the worms; making cheese and yogurt, and...) puts us in the position of necessarily understanding interdependencies and developing a working knowledge of nutrition, health and production, seasonal weather influences, reproduction, animal behavior (human included), uses of herbal preventatives and remedies, and advantages and problems of domesticating animals. Also acquired in the process are skills in the planning and construction of sheds and fences, making dairy products (butter, yogurt, kefir and farmer's cheeses, not to mention custard and ice cream) and meat-animal raising and butchering (pigs, chickens, beef, rabbits). To leave out the "find-the-hole-in-the-fence and put-the-cows-back in" would lessen the validity of this account. To a cow the grass is always greener..., and to a

Figure 5 — Six-thirty am

cow's caretaker it means a never-ending task of mending fences. As far as animal health care, the responsibility of two cows, two calves, chickens, rabbits and pigs means discussions with vets, various experts, local farmers and library resources to provide experienced, specific, and current information and ideas.

The development of the livestock program depends in part on the simultaneous development of agriculture (lack of water being a major problem at present). If we produce more of our own food for animals we could increase the kinds and numbers of animals. Finding the balance between food potential (plant and animal) of the land, and food needs of livestock and people, is a constant process. The balance varies from season to season, session to midsession, and with the evolving focus of the Farallones Rural Center.

Educational Program at the Rural Center
The emphasis at the Rural Center has been the development of an apprenticeship program to train skilled technicians in several key areas such as land management and agriculture, solar systems, and small-scale on-site waste management systems. Besides providing training in general vocational skills, the Institute's program attempts to provide a broad array of learning experiences in a real life application through a live-in, hands-on educational experience. At the present time, for example, there is an extreme shortage of skilled technicians in the solar energy field that can provide solar expertise to householders and others at an accessible cost. Our programs combine basic theoretical and engineering background with a hands-on skills mastery that gives the graduate solar technician the ability to enter the job marketplace in a position to succeed financially and to offer viable energy alternatives to homeowners.

The educational program at the Rural Center is two-pronged; we provide the opportunity for students (who may receive college credit through our affiliation with Antioch College/West) to

Solar Architecture

come and participate in our whole-life systems residential program which runs for 10 weeks three times a year. We also have an apprenticeship program for advanced students who live at the Center for a full year, working closely with skilled staff people and involving themselves in money-producing projects outside the Institute (for example, two apprentices will begin construction shortly on a Farallones designed solar-heated home up the road).

Our Whole Life Systems program is open to all ages, and our past students have included architects, engineers, a medical doctor, college and high school students. Our educational program is based on the belief that people learn by doing rather than be being talked to, so the experience at the Rural Center is a work one rather than an academic one. Students choose a major area of concentration within the integrated areas of energy systems, waste recovery, food production and shelter construction, and work primarily in that field under the direction of a staff person. While the major emphasis of the Farallones program is work-oriented, the Institute also provides some classroom time, and there are weekly basic skills workshops in a diverse number of activities such as blacksmithing, electrical layout, drafting and automobile mechanics. In addition, the Institute provides the opportunity for students to interact with skilled outside resource persons during hands-on workshops, seminars and discussion.

FARALLONES URBAN CENTER: THE INTEGRAL URBAN HOUSE
In October of 1974 the Institute purchased an old, delapidated Victorian home on an 1/8-acre city lot in Berkeley. Our intention was to transform the weathered, abandoned dwelling into a model of an ecologically sound urban habitat. Accommodations were made in the building's structure to make the most efficient use of solar energy for heating and cooling the dwelling. The plumbing installation was designed to facilitate complete recycling of household waste water and human excrement right on the premises. The yard was landscaped with fruit trees, food-producing ornamental foliage, and a large vegetable garden.

Facilities for the raising of fish, rabbits and chickens were constructed. In short, the entire lot was designed to promote the maximum degree of self-reliance possible on a standard size city lot. The Urban House serves as a model for a more ecologically sound urban habitat, and to provide urban dwellers with physical and conceptual tools for creating a more self-reliant life style.

Presently four people live in the house. All life-support systems are in full operation. The household water is heated by a solar collector. "Passive" systems of solar capture are employed to regulate internal house temperatures. All the vegetables and meats which the residents consume are produced on site. The household waste water and sewage is recycled via a series of biological treatments and used to fertilize the vegetable garden and dwarf orchard. And, surprisingly, the operation of the household requires only eight hours of labor per week. Design of the home's life-support systems was conducted with the objective of providing the greatest degree of sustenance for the least amount of labor input.

Space Heating
In addition to exceptionally good insulation, the Integral House has several features which contribute to its space heating. For instance: when the old Victorian mansion was rebuilt by its present owners, it was fitted with a large number of south-facing windows and only a few windows on the north side. This produces some degree of passive solar heating during the day. (The upstairs windows are equipped with insulated shutters that are closed to help retain warmth at night.)

Within the building's bathroom is a smaller version of Steve Baer's Drum Wall: a bottle wall, in which one-gallon glass jugs - filled with ink-blackened water and supported in rows just inside the windows - serve as heat sinks that absorb the energy in the sun's rays. At night, insulated shutters (outside the windows) can be closed to keep the bottles' stored heat inside where it radiates into the room.

Solar Architecture   325

Much of the IUH cooking is done on a beautiful old combination
gas and wood stove (equipped with an O-shaped - or doughnut -
stovepipe that radiates into the room a bit more of the heat
which would otherwise go up the chimney). The house also uses
a large solar oven - built by a student - for baking casseroles
or bread on sunny days.

Solar Heated Water
Nearly all the IUH's hot water needs are met by a solar heater
that can warm 120 gallons of water in an attic storage tank to
surprisingly high temperatures...often past 170 F. The 86-square
foot solar collector that is the heart of the system is the
"Ritz of homebuilt collectors" and is expected to last the
lifetime of the house. Doug Daniels, who helped design the
solar-powered heater, says the complete system - including
storage tank and pipes (items that would have been required for
a conventional electrically operated heater anyway) cost around
$9000 to build, not counting labor. Taking into account the
utility bills and maintenance costs associated with more traditional water heaters, Doug figures that the solar-powered
setup should pay for itself in ten years.

A small electric water heater acts as an emergency backup
system. (Although electric water heaters are less efficient
than gas-fired units, the pilot light on a gas heater must burn continuously. And, since the IUH backup is called upon so infrequently, that would be rather wasteful.) So far, the electric standby has been used only three times...and even then it was relatively efficient

| Figure 6 | Collector panels for solar hot water heating system |

since it was being fed water that'd been pre-warmed to 95 F by the solar heating system.

Urban Gardening
Even in the dead of winter the IUH garden is lush with foliage and brimming with vegetables. Because the size of the lot precludes the planting of long rows, crops are sown in raised beds that surround the house. (Plant varieties are rotated from bed to bed to keep specific soil nutrients from becoming exhausted in any one section of the garden, and seedlings are grown in the greenhouse so that the beds are always occupied by mature - or nearly mature - plants.) Small avocado, fig and quince trees stand above raised beds closely planted in potatoes, broccoli, lettuce, tomatoes, corn, peas, beets, carrots, celery, spinach, chard and squash. Salad greens, scallions and herbs are grown on the porch (adjacent to the kitchen), while nearby are perennial patches of strawberries, rhubarb, and asparagus. In addition, dwarf fruit trees - espaliered to the north wall of the house - will soon provide lemons, plums and three kinds of apples.

No chemical fertilizers are used on the garden. Rather, a one-inch-deep layer of compost made from kitchen garbage, rabbit manure, grass clippings, sawdust and other wastes is maintained on the garden's beds to (1) act as a mulch which keeps weeds down and (2) make the soil light, airy and rich in nutrients. (No tilling is ever needed.)

Because of the rich diversity of plantings in the garden, insects rarely pose a problem. (Small plantings of many types of crops tends to prevent mass infestation by any one kind of pest.) And when insects do pose a problem, biological controls - such as natural insect predators and specialized diseases that affect only the pest in question - soon "settle the hash" of the unwanted intruders.

To further make the point that _anyone_ - even apartment dwellers with no access to cultivatable land - can grow their own food.

Solar Architecture

Integral House residents have created a rooftop garden of containers filled with pure compost. (The compost is not only rich in plant nutrients, but is lighter than soil and thus lessens the load that would otherwise be placed on the building's rafters.)

Animal Husbandry

The Integral House's food-raising efforts extend not only to the growing of fruits and vegetables but to the production of animal protein as well. The latter, in this case, means chickens and rabbits.

All together, about 15 chickens - layers and fryers - inhabit the Institute's urban homestead. Four hens live in a "composting house" on the roof, where - in addition to laying eggs within ten feet of the breakfast table - the birds produce rich manure for the compost heaps. The remaining chickens are kept at ground level, on the shady north side of the old Victorian building.

Some 10 to 20 rabbits, depending on whether or not a litter has been born recently, are also housed on the structure's shady side. Commercial pellets, garden-grown alfalfa, and discarded produce obtained from a nearby market make up their diet.

In addition to chickens and rabbits, the Urban House has an aquaculture program designed to determine if the production of fish and crustaceans can be made feasible for city dwellers. Sacramento blackfish and rainbow trout are being raised in an experimental fish pond in the house's small yard.

Along with the daphnia and algae that grow naturally in the pond, the "livestock" feed on worms and bees raised by the House. The worms are grown in sawdust-covered trays mounted below IUH chicken cages to catch the birds' droppings. (Which thus serve as both fish food and "workers" that speed the production of compost from the chicken manure.)

The bees, the hives of which are located above the pond, fall into the pond only occasionally, and by accident, as they return to either of two hives located above the body of water. Says Sterling: "Happily, the hives contain so many bees that the loss - now and then - of a few unlucky ones doesn't hurt anything."

At present, Bunnell is installing a biological filtration system designed to remove growth-inhibiting wastes (produced by the fish) from the pond's water. The system is stunningly simple: It's nothing more than a bed of oyster shells - coated with bacteria that feed upon (and filter out) impurities in the water - through which the body of water's effluent is passed. If it works, the filtering system should significantly increase the yield (by weight) of fish from the small pond.

Educational Program
An assortment of educational programs geared to satisfy a wide range of needs is offered at the Urban House. Public classes are held on a monthly basis in the fields of solar energy systems, habitat design, urban food production and waste recycling. "Hands-on" workshops are conducted in small stock raising beekeeping and aquaculture. Apprenticeships are available for people who would like to study for longer periods. Teacher-training programs and environmental education seminars are conducted in the development of instructional skills for classroom teachers. Professional consultation sessions are offered for people seeking guidance in modifying their own houses and public tours of the project are conducted on a regular basis.

BIOGRAPHICAL SKETCH
Alison Dykstra is an architectural designer and Acting Director of the Farallones Institute/Rural Center in Occidental, California.

INDEX

Abu Simbel.................6
Anaerobic Digestion........257
Analytical Model...........90
Appropriate Technology.....307
Archeostronomy.............2
ASHRAE.....................118
Aspen Community School.....292
Autonomy...................12
Battery Storage............237
Beta Project...............301
Bio-generation System (BGS).267
Bottle Wall................324
Caracol....................8
Chaco Canyon...........9-135
Chichen Itza...............8
Climate........115,153,154,157
Climatic Envelope......103,104
Clo........................42
Collectors..152,163,191,205,215
Comm. Col. of Denver.13,201-206

Composting Toilets....271,314
Computer Simulation.....66,92
Conduction...........116,138
Craven Residence..........180
Decade.................12,80
Design Temperatures.......116
Direct Gain............62,310
Dry Toilets....14,185-187,273
William Edmondson.133,134,139
Elec. Solar Sys...161,165,167
Environmental Ethic.......295
Eutectic Salt.........173-177
Farallones Institute......307
Farallones Urban Center...323
f-chart...............150,151
Flush Toilet..............272
Focusing Collector........155
Franta Residence......186,187
Gallium Arsenide Cells....162
Gaudix, Spain 29

Geothermal Energy.......249-253
Greenhouse......65,101,107,119,
                123,133,143,312
Heat Exchanger..............144
Heat Mirror............101,102
Heat Pump..............147,206
Hothouse....................193
Human Heat Transfer..........50
Infiltration...........116,117
Insolation.........139,153,171
Jane Jacobs..................36
Lazy-O Ranch................197
Meadowood Apartments....180,181
Mesophilic Range............258
Methane...........182,255,262
Microclimate.................23
Mollica Residence.......184,185
Movable Insulation..........107
Northern Windows............111
Optical Shutter.............101
Orr Greenhouse..............133
Passive Solar Systems.59,89,101
P.E.I. Ark...................13
pH......................256,258
Photovoltaic.......161,162,164
Pitkin Cnty Air Term....182,183
Pitkin Cnty Bus Stops.......184
Place Bonaventure............29
Place Ville-Marie............29
Plant Growth................131
Protein.....................263
Provident House..............12
Pueblo Bonito..............9,35
Ramses II.....................6
Ray Conservatory............189
Roaring Fork Resource Center...
                179,287-290
Roof Pond...................310
RTD-Denver...................13
R-value.....................116
St. Benedict's Mon....181,182
SCP.........................263
Shore Residence.......182,183
Silicon Cell................162
Sky Therm....................66
Smith-Hite Studio.........185
SNG.........................262
Sod Roofs.................27,33
Solar Dome..................292
Solar Electric....209,214,223
Solar Gain...................90
Solar Hot Water.............312
Solar One Holer™.........275
Solar Power Plant.........209
Solar Sustenance Project..119
Solar Trough...............196
Solar Wall.................103
Soltran™..................275
Specie Mesa House.........190
Stonehenge...................2
Sundown.....................25
Sun-louvers................186
Sunshine Mesa Wind Inst...190
Karen Terry Residence......11
Thermal Storage....90,93,127,
        129,133,146,159,173,220
Thermal Storage Mater...97,98
Thermocrete................101
Thermosiphon................66
Wendall Thomas..............32
Transparent Insulation....101
Trombe Wall.............64,91
Urban Gardening..........326
U Value................93,115

U Factor....................140
Van Winkle Greenhouse.......143
Vegetation...................23
Ventilation........124,125,130
Wallasey School..............61
Water Reclamation...........318
Water Wall...............63,324
Wet-Dirt Thermal Storage....133
Wildwood School.............295
Wind Power..................227
Wind Systems................314
Winterstash Greenhouse..198,199
Wood Stoves.....185,186,241-246
Wyon, Daniel P. 48

More Energy Books From
ANN ARBOR SCIENCE

SOLAR ENERGY: TECHNOLOGY AND APPLICATIONS (Revised Edition) — Williams
The Consumer's ELECTRIC CAR — Wakefield
1977 SOLAR ENERGY & RESEARCH DIRECTORY
EXTRACTION OF MINERALS AND ENERGY — Deju
COMBUSTION: FORMATION AND EMISSION OF TRACE SPECIES — Edwards
ENVIRONMENTAL ASPECTS OF NUCLEAR POWER — Eichholz
FUTURE ENERGY ALTERNATIVES (Revised Edition)—Meador
PERSPECTIVES ON THE ENERGY CRISIS, Vol. 1 — Gordon/Meador
PERSPECTIVES ON THE ENERGY CRISIS, Vol. 2 — Gordon/Meador
THE CONCEPT OF ENERGY — Hoffman
FUEL AND THE ENVIRONMENT — Institute of Fuel
THE UNSETTLED EARTH — Jones
FUSION REACTOR PHYSICS — Kammash
POWER GENERATION: AIR POLLUTION MONITORING AND CONTROL — Noll/Davis
EMISSIONS FROM COMBUSTION ENGINES AND THEIR CONTROL — Patterson/Henein
SOLAR DIRECTORY — Pesko
ENERGY, AGRICULTURE AND WASTE MANAGEMENT — Jewell
FUELS, MINERALS & HUMAN SURVIVAL — Reed
INTRODUCTION TO ENERGY TECHNOLOGY — Shepard
TECHNOLOGY AND HUMAN VALUES — Watkins/Meador
THERMAL PROCESSING OF MINERAL SOLID WASTE FOR RESOURCE AND ENERGY RECOVERY — Weinstein/Toro
SCIENCE AND TECHNOLOGY OF OIL SHALE — Yen

Discarded Date JUN 25 2025